"Who is killing the Colorado River? In this environmental detective story, Stephen Grace points fingers and names names, but, best of all, he finds some hope and possible solutions."

—Mark Obmascik
Pulitzer Prize-winning journalist and bestselling author of *The Big Year* and *Halfway to Heaven,* winner of the National Outdoor Book Award

"An oilman who's also a conservationist? Who knew?! But that describes Bud Isaacs, the inspiration behind this book, and it's his energy and expertise that might save the West's critical upper Colorado River."

—Greg Dobbs
Emmy Award-winning TV network journalist

"A great read about the struggle to preserve a precious piece of the nation's natural heritage and the difference one concerned and very determined citizen can make. Although the story focuses on the headwaters of the Colorado River, it also offers valuable lessons for how diverse interests can, eventually, come together to solve water disputes throughout the American West."

—Fritz Holleman
Former Deputy Solicitor for Water Resources,
U.S. Department of the Interior

"Most of us don't think much about it when we turn on the tap to get a glass of water. In this terrific story, Stephen Grace shows us some of the complex issues and fascinating characters that are behind every drop we drink in Colorado and the West."

—Anders Halverson
National Outdoor Book Award-winning author of
An Entirely Synthetic Fish

"In *Oil and Water,* Stephen Grace gives voice to an unlikely cast of characters whose words weave a compelling tale of the death and potential rebirth of one of America's iconic rivers, the Upper Colorado. With the grand sweep of a James Michener novel, the clear-eyed reporting of John McPhee's work, and Grace's own personal insights, *Oil and Water* deserves a place with the classics of environmental literature."

—Susan J. Tweit
Plant biologist and award-winning author of
Walking Nature Home: A Life's Journey

"At its root, conservation is about people. In a conversational and easy-to-read style, Stephen Grace shows us in *Oil and Water* how one man can make a big difference in helping to protect the lands and waters that sustain us all."

—Chris Wood
President and CEO of Trout Unlimited

PRAISE FOR STEPHEN GRACE

Under Cottonwoods

"In his first novel, Grace writes with a lyrical power, celebrating the healing power of the human spirit set free in the wilds."
—*Los Angeles Times*

"A humane and hopeful story that will engage both nature lovers and champions of the human spirit."
—Wally Lamb
New York Times bestselling author of
She's Come Undone and *I Know This Much Is True*

"This first novel is so splendidly crafted and realized, it shines."
—Stratis Haviaras
Founder of *Harvard Review*

It Happened in Denver

"By the evidence of this excitement-packed narrative history, Denver has been simply bubbling with major events."
—Barnes & Noble

Colorado: *Mapping the Centennial State through History: Rare and Unusual Maps from the Library of Congress*

"An excellent and unique addition to the vast library of works on Colorado history."
—Center for Colorado & the West at Auraria Library

Shanghai: Life, Love and Infrastructure in China's City of the Future

"Stephen Grace has a love affair with Shanghai, and he portrays it wonderfully."
—Mel Gurtov
Editor-in-Chief, *Asian Perspective*

"Grace understands his place in the story and his are the novelist's sensibilities; his observations and scenes are in turn horrifying and hilarious."
—Jon Billman
Author of *When We Were Wolves*

Dam Nation: How Water Shaped the West and Will Determine Its Future

"Grace acts as both poet of Western wilderness and a knowledgeable translator of water policy."
—*High Country News*

"Grace…effectively, even humorously at times, captures the highlights of the West's liquid history."
—*Kirkus Reviews*

"Grace covers an ambitious amount of material, yet manages to draw in the reader with colorful and engaging stories that reveal a deep connection with the subject."
—*University of Denver Water Law Review*

"The words 'lyrical prose' and 'water policy' seldom are written in the same sentence, but author Stephen Grace has melded the two in his new book, *Dam Nation*."
—*Steamboat Today*

Grow: Stories from the Urban Food Movement

"*Grow* is gorgeously written and a true pleasure to read."
—Helen Thorpe
Author of *Just Like Us* and *Soldier Girls*

"Stephen Grace has written with passion, wisdom, and—yes—grace about the backstory of the food on our plates, and about the people in Denver who are working to bring that story closer to home."
—Nick Arvin
Author of *The Reconstructionist* and *Articles of War*

"A captivating and original book.... A must-read—for everyone."
—Laura Pritchett
PEN USA Award-winning author of
Hell's Bottom, Colorado

The Great Divide

"Trust Stephen Grace to write about complex matters with clarity, fairness, and (there is no avoiding the word!) grace, and, at the same time, prepare to benefit from remarkable, thought-provoking photographs."
—Patty Limerick
Faculty Director and Board Chair,
Center of the American West

"A fluid, eloquent, painstakingly researched and deeply intelligent rendition of one of the most complex and far-reaching subjects in the West."
—Kevin Fedarko
National Outdoor Book Award-winning
author of *The Emerald Mile*

"A powerful journey with evocative visuals through the history and current state of affairs of western water: highlighting how we use it, cherish it, and abuse it. A must read for any Westerner."
—Pete McBride
Author, Filmmaker and Photographer,
National Geographic

"A must read for anyone interested in the history and future of water in Colorado."
—Ken Salazar
Former U.S. Senator from Colorado,
Former U.S. Secretary of the Interior

"Grace possesses deep insight and a strong sense of place; this presentation, coupled with [Jim] Havey's remarkable photos and occasional archival images, is exceptional."
—*Publishers Weekly*, Starred Review

OIL and WATER

*An Oilman's Quest to Save the Source of
America's Most Endangered River*

Stephen Grace

Published in the United States by

UCRA Publishing
1600 Broadway
Denver, Colorado 80202
www.UCRA.us

OIL AND WATER

SEVENTY-FIVE PERCENT of the proceeds from this book will go to the Upper Colorado River Alliance to be used for the continuing restoration of the upper Colorado River watershed.
www.UCRA.us

ONE OF THE most overused and fought-over natural resources in the world, the Colorado River provides drinking water for nearly forty million people in the United States and Mexico, generates hydroelectricity that powers cities, irrigates four million acres of farmland, supports twenty-two Native American tribes, and maintains a multibillion-dollar tourism and recreation economy. Seven national wildlife refuges, four national recreation areas, and eleven national parks depend on the river. The natural wealth nourished by the Colorado is incalculable—and is at dire risk of disappearing.

A SYSTEM OF dams, reservoirs, pipelines, pumps, and tunnels redirects water from the upper Colorado River watershed eastward to the farms and faucets of the Front Range. Diversions across the Continental Divide have enriched cities and agriculture on the eastern plains while impoverishing the ecosystem of the Colorado River headwaters.

CONTENTS

In an age when man has forgotten his origins and is blind even to his most essential needs for survival, water along with other resources has become the victim of his indifference.

—Rachel Carson, *Silent Spring*

AUTHOR'S NOTE

In the summer of 2014, I was invited back to Brazil, where the previous year I had helped launch a cultural exchange program for young American leaders. I decided not to return. Oilman Bud Isaacs convinced me that instead of running off to the Amazon to help Brazilians become better stewards of their natural resources, I should focus on a battle unfolding in my own backyard.

In 2013, after publishing *Dam Nation: How Water Shaped the West and Will Determine Its Future*, I had thought I was finished with the topic of water in the American West. Tasked by a publisher with writing a book that would introduce the layperson to the arcane world of western water, I had wandered into the topic as a complete beginner. I immersed myself in it, drowned in it, wrote many bad puns about it, and struggled to make the subject accessible. Of all the topics I had tackled, including Chinese language and atomic clocks, water was the most complicated.

I focused on the conflict at the heart of water scarcity in the West—in large part to keep readers from slipping into comas.

Friends asked me over beers the subject of my next book. Maybe a primer on watching grass grow? An introduction to seeing paint dry? I pointed out that the most important ingredient in the beer we were drinking was not barley or hops but...wait for it...*water*! They were unimpressed. I insisted water was everywhere—even in the smartphones they checked while I babbled like a mountain brook about the importance of water to our lives. I explained that the data centers that enable Google searches for the phrase "water crisis" use massive amounts of water for cooling, and the silicon chips that run our smartphones require purified water in their manufacturing process. Water, water, everywhere: I made my case to friends and strangers. Some listened politely; many drifted toward sleep.

I am a writer largely because there's not a snowball's chance in Albuquerque of me thriving in the real world. I'm grateful I didn't go to law school and study water law, as so many people in Colorado have done. That "water attorney" is a popular vocation in the state has provoked laughter from my buddies in Boston and Manhattan. I warned them that if Westerners didn't get their water crisis under control, there might not be water available for snowmaking—which, combined with climate change, could put the kibosh on ski trips in the Rockies. This potential lack of powder gave them pause. But then they harassed me about wasting my time with water. And I wondered if they weren't right.

"Non-Tributary Groundwater" differs from "Tributary Groundwater," which, believe it or not, is different from something called "Not Non-Tributary Groundwater." Compared to western water law, calculus is a cinch, Mandarin a breeze. When I spoke with people working in the wonky world of water, sentences like this one battered my brain: "There were deleterious ancillary effects to drawing down the alluvial aquifers." This did me no good in my ongoing struggle to maintain an attention span longer than that of my Labrador.

So in 2013, after publishing *Dam Nation*, I had boxed up my water books and headed to Brazil to help the best and brightest culled from America's colleges and workplaces teach

English in a rising nation with the seventh-largest economy on Earth—a country tearing through its natural resource base at a frightening pace. Through cultural exchange, both parties would benefit, and we would all march hand in hand into a new world of international understanding and cooperation. That was the idea, anyway.

Many selfies were shot, many Facebook posts were made with said selfies, and the ideas discussed often revolved around episodes of *Breaking Bad* that the best and brightest streamed on their tablets. It is unclear to me whether the young Americans I accompanied to Brazil were aware they were actually in Brazil and not participating in a virtual reality scenario on a screen.

The greatest achievement of our program in Brazil was, arguably, teaching the "Cha Cha Slide." If you aren't familiar with the dance steps, well, I envy you. One day as I jumped and I hopped, and a world away the Colorado River spiraled toward terminal decline, I did some soul-searching. I wondered why on earth I was in the Amazon Basin doing the Cha Cha Slide.

An American leader of the program told me she didn't understand what carbon had to do with climate change. More surreal than following a shaman through the rainforest was watching scientifically challenged Americans serve as cultural ambassadors. High school students in the Amazon were eager to explain the carbon cycle and Brazil's pioneering use of alternative fuels. I gave Brazilian students English lessons; they gave me hope for the future. I encountered hundreds of young Brazilians so earnest and determined to lift themselves and their country with their hard work and creativity, they reminded me of World War II Americans—I met Brazil's greatest generation. Humbled, I returned home.

Instead of rushing off to other countries to spread our wisdom while we stream TV shows and post smiley face icons on Facebook, perhaps we should take a look in our own backyard and scrutinize our stewardship of the natural resources entrusted to us. My Brazilian friends can decide for themselves what to do with their Amazon. We need to figure out how to save our Colorado River.

In my backyard is a river in danger of dying. As the state of Colorado's population doubles in coming decades, pressure will mount to divert what remains of the headwaters to faucets and lawns on the crowded Front Range. The source of this great river could be sucked dry.

By fighting to preserve the Colorado River in the wealthiest society the world has ever known, people like Bud Isaacs are telling a story as important as the struggle to keep the waters of the Amazon flowing free through primeval forest. If we can't stop killing the Colorado, we will have a lot of explaining to do in Brazil, and all over the world, and to future generations. How we manage the Colorado River is a test. And much depends on whether we succeed or fail—from our willingness to embrace science, to our ability to curb rapacious consumption, to our resolve for safeguarding resources exquisite and rare.

MY FAMILY HAD LITTLE MONEY when I was growing up and we seldom traveled beyond the state borders of Missouri. A trip two hours south of our suburban home in St. Louis to explore the Ozarks was high adventure in my childhood. Spring-fed streams that spilled from limestone bluffs held in their cool waters sunfish and crawdads and burbled in hot summer nights beneath luminous swarms of fireflies. In the woods of Missouri I became a lifelong naturalist obsessed with rivers and streams.

One summer my family left St. Louis in a mechanically suspect sedan on a road trip to Winter Park, Colorado, where we bounced some checks and rented a ski condo for off-season rates. I waded into the chilly riffles of a mountain stream wearing soccer shoes and carrying a spinning rod from Kmart. I was a day's drive from Missouri, but I might have been in Tibet. The unblemished blue sky, the soaring snow peaks: I glimpsed a higher kingdom that had been hidden from my eyes by Missouri's humid skies.

I stood in the Colorado River headwaters with my feet turning numb, surrounded by undercut banks woven thick with willows and filled with clear-flowing water that turned to jade in the deepest pools. The silvery flash of a fish struck my mind like a quick thought I couldn't quite grasp. I was instantly addicted to the rivers of the West—and I have been ever since. Now I have been to Tibet and many other memorable places on the planet, but the upper Colorado River watershed is as magnificent as anything I've seen.

A few years after first wading into the rivers of Winter Park, I became what today is called an "at-risk youth." An outdoor adventure program in St. Louis sent me West, away from my poor decisions and into the mountains of Colorado. I stood in the moving waters of a cold river that had carved canyons the color of blood, and I realized I needed to change. While writing this book, I learned my best friend from high school died from liver failure caused by alcohol and heroin abuse. I had fun kicking soccer balls on the green grass of my Missouri childhood, but the rivers of the West saved my life. Killing the Colorado so suburbs, no different from the sprawl in St. Louis where I grew up, can have lush lawns and shopping malls, and so industrial farms can grow commodity crops, is unacceptable.

I approached this book with the tools of journalism—but total objectivity was not among them. I am biased in favor of saving the river that once saved me. That said, I have tried to treat everyone in this book with fairness. I am not on any person's side. I am on the side of the river.

PROLOGUE
THE RIVER

Whenever water flows from a faucet in Denver, whenever a head of lettuce is harvested in the Imperial Valley of California, the Colorado River moves closer to becoming a dry bed of stones. Most years this exhausted watercourse is drained before it reaches the sea. In the days before the river was dammed and diverted, the green lagoons of its delta seethed with fish and its skies dimmed when birds blocked the sun. Now the Colorado ends in a plain of mud that fractures into countless pieces—a fitting end for a river over which every drop is fought so fiercely, civilization itself could crack.

In the summer of 2014, as drought descended on the Southwest, Lake Mead sank to its lowest level since it started to fill with Colorado River water behind Hoover Dam in the 1930s. While the national media rushed to the desert to report on water spiraling down the drain of the nation's largest reservoir, a study by NASA and other researchers revealed what was going on beneath the surface of the Colorado River Basin. More than

75 percent of the water lost in the drought-ravaged region in the past decade had been extracted from underground. The Colorado River forms the tip of the iceberg in a water cycle that largely lies hidden below. The NASA study revealed that both the tip and the berg were vanishing at a frightening rate as cities and farms sucked water from beneath the earth.

Suddenly the delta's fate in Mexico seemed not an isolated tragedy for a distant reach of river but a harbinger of things to come along the Colorado's winding course. Lights turning off in Las Vegas; residents of Phoenix fleeing water pipes clogged with dust: These scenarios started to seem less like the stuff of apocalyptic fiction than actual consequences of drought. As conditions in California worsened and voluntary water restrictions failed, the state took the unprecedented step of instituting mandatory water restrictions to save precious drops. Water agencies policed the streets, punishing people who washed their cars with hoses that didn't have shutoff nozzles.

But the water crisis evolving in cities supported by the Colorado River is about more than carelessly washing cars and growing lush lawns in the desert—practices that will, of necessity, soon disappear. Lake Mead, circled by mineral rings like an emptied bathtub, is the largest water source for a region that runs one of the world's most efficient economic engines. And the natural wealth supported by the Colorado is beyond measure. All the creatures that depend on its flow, including humans, are in danger as this dammed and diverted river runs dry.

There are, however, signs of hope.

In the spring of 2014, while shrinking reservoirs in the Colorado River Basin posed an existential threat to cities and agriculture across the western United States, an effort to resurrect the dead delta in Mexico got underway. From Morelos Dam on the U.S.-Mexico border, a "pulse flow" was sent slipping past the structures of concrete and steel that block the Colorado's progress toward the Pacific. And for a moment, the timeless cycle was restored.

When the Colorado once again met the salty bulk of the ocean, the vaquita, the smallest sea mammal on the planet, and arguably the cutest, was given a temporary reprieve as a brief blessing of freshwater infused the Sea of Cortez. *Vaquita* in Spanish means "little cow." Dark rings surround the eyes of this tiny porpoise with large fins that slice the water as it forages for fish in shallow lagoons once nourished by nutrients carried by the Colorado. Nurseries that half a century ago produced vast clouds of shrimp and schools of fish so dense they packed the river's estuary have vanished. The vaquita is one of many forms of life—from humpback chubs to Angelinos, from dippers to Denverites—that depend on the Colorado River. And all these creatures have a precarious hold on a rapidly changing planet.

In one human lifetime, an area the size of Rhode Island has been transformed from a river swollen enough to float steamboats to a plain of parched and barren sand. But when a small amount of water was allowed to escape the dams and diversions upstream of the delta, signs of life soon emerged. As the river was freed from its long captivity in the reservoirs of the north, waterfowl filled the sloughs and backwaters and mirrored ponds. And an explosion of human energy reflected the riot of plant and animal life in the delta.

Residents of San Luis Río Colorado performed a community trash cleanup to prepare the riverbed for the coming water. People picnicked and partied along shores that had crumbled in dryness for decades. When the Colorado touched the tides of the sea, citizens of two nations celebrated.

But while the media focused on this feel-good story of rebirth in the delta, the headwaters region more than a thousand miles upstream was slipping toward death.

LIKE WATCHING A FILM run in reverse, we can follow the course of the Colorado from its abrupt end in its delta back to its subtle birth in mountain snows. Starting at sea level at the Gulf of California, we rise more than 14,000 feet into oxygen-thin air toward the upper ramparts of the Rocky Mountains. Over the 1,450-mile length of this odyssey through the dry heart of the Southwest, we pass the All-American Canal, an aqueduct in the Sonoran Desert of California that transformed a tawny land of sand dunes sketched by sidewinders into a green patchwork of alfalfa fields. Further upstream, Lake Mead shimmers in the desert, a vast reservoir that seems the size of a sea. From its impounded waters the city of Las Vegas drinks.

Before the Colorado was forced to settle behind dams as tall as skyscrapers, the crashing torrent of its flow trenched downward through deep time. The Grand Canyon's walls show evidence of 1.8 billion years of change, beginning with ancient crust that cooled atop the molten core of the world, and then rising through layers of magma and marine life, through strata of cinder and sand. As the region shifted back and forth between deserts and shallow inland seas, a strange zoo of fossilized life was frozen in stone. Many wondrous forms were preserved, from trilobites, the first creatures with eyes to see the world, to animals known as sea lilies that look like flowers. The footprints of scorpions and lizards that scurried across ancient dunes were seized in rock. Silts and sands rich in iron squeezed together and stacked up in stone bands painted in shades of red through the Permian, the Triassic, and the Jurassic periods. The great upheaval that raised the Rocky Mountains also pushed the Colorado Plateau skyward. Rivers gathered across the tilted land. And water, in its infinite persistence, sliced the solid earth.

Over six million years the Colorado River pulverized a vertical mile of stone and moved this sediment downstream to the Sea of Cortez, where it piled up in a delta as deep as the highest mountain in North America is tall. The debris spread fanwise into the sea, forming the lush lagoons that gave sanctuary to vaquitas and jaguars, to beavers and birds.

Upstream from the Grand Canyon, the great curved face of Glen Canyon Dam calms the Colorado. Behind this concrete slab, Lake Powell backs up, its stilled water spreading through canyonlands in a network of branching arms. Warm water red as blood pours into the deepness of the reservoir, depositing many tons of silt as it stalls and cools. When released from its confines behind the dam, the water is chilly and green—more like a glacial stream than a desert river.

Tributaries of the Colorado River helped sculpt the Colorado Plateau, and life evolved based on the seasonal rise and fall of their flow. The Yampa is the last of these wild rivers that rage in spring and then relax to lazy currents in late summer. Largely undammed and undiverted, the Yampa is a reminder of what the Colorado River Basin once was. The Yampa still moves as it has for millions of years from the white snows of Colorado's Flat Tops Wilderness to the muddy chaos of its convergence with the Green River.

In the waters of the Yampa and Green, the Colorado pikeminnow grew up to six feet long. This top predator that finds its way through murky currents back to where it was born once hunted and spawned in the rivers of the Colorado Plateau in such abundance that a commercial fishery canned the creature and sold it as "white salmon." Now the last of this species that evolved in the silt-filled floods of the Colorado River Basin survive in a few reaches unaltered by dams and diversions.

As we return from our tributary detour to the mainstem Colorado River and continue our travels upstream, we enter Gore Canyon, where the pre-dammed Colorado cut a narrow slot through Precambrian granite. Walls painted with pastels and the darkness of igneous rock rise a thousand feet on both sides of the river, squeezing the flow to thunder and spray.

When we make our way toward the Colorado's source, the surging river that carved the canyons of the Southwest quiets to a trout stream that meanders through hay meadows. As we scale peaks closer to the river's birthplace in Rocky Mountain National Park, we find trickles clear as tap water and a few degrees removed from ice. Climbing higher yet, we reach summits of the Great Divide, the crest of the continent, where snow is cached each winter and waters part ways as they travel toward one of two oceans on our blue planet. Melted snow from the Continental Divide finds its way down the West Slope and the East Slope of the Rocky Mountains by nature's design—or by human diversions that can drain the river till it runs dry.

THE TALE OF THE COLORADO River's dead delta and its rebirth
with a pulse flow was covered extensively by the media. Many
people now know the river's plight downstream—images of
the stark sands of the delta stretching into the distance are
emblazoned in our collective consciousness. But few know the
story of the river running dry near its source, the stones of its
emptied bed steaming in the alpine sun of a mountain valley.

On Labor Day in 2006, a foreman at a ranch outside Kremmling, Colorado, opened a headgate and dried up the Colorado River. The upper reach was so overused that summer, a
single diversion to satisfy a ranch's water right was the thou-
sandth cut that caused the river's death.[1] The municipal water
provider in Denver, which sits on the other side of the mountains but drinks deeply from the Colorado River's source, gave
the corpse of the Colorado an emergency cure by releasing
some flow from one of its reservoirs. The river was temporarily revived; immediate crisis was averted. But the headwaters
to this day linger on life support. And the prognosis for the
source of the Colorado is every bit as dire as that of its delta
downstream.

Along with supporting cities, industry, energy, and agriculture, this river that is worked to the bone is relied upon by ecosystems and recreation in national wildlife refuges, national
recreation areas, and national parks. But nature spills beyond
the boundaries of protected areas marked on maps to encompass every living thing.

Along a stretch of the Colorado River below a reservoir
known as Windy Gap, one of the West's most prolific trout
streams is in precipitous decline due to a small but destructive dam and too many diversions. For more than a dozen
river miles downstream of Windy Gap Dam, the Colorado is
drained dangerously low before creeks and rivers reinvigorate

---

1   This troubling event was reported by Mark Jaffe in a 2008
Denver Post article, "Current Affairs on State Water."

its flow. This forgotten stretch below Windy Gap is known as "the hole in the river." And into this hole an oilman has poured his heart.

~~~~~~~~

As a LIFELONG OILMAN, Vernon "Bud" A. Isaacs Jr. may seem an unlikely person to lead the battle to preserve an endangered river. I was drawn to his story because he has been on the receiving end of plenty of wrath from environmentalists like me for his work in oil and gas. But for fifteen years he has been broadcasting the wrongs done to the upper Colorado River and working to make things right. This struck me as flat-out strange. *An oilman who wants to save a river?*

Bud listens to Fox News on the stereo in his shiny new Porsche; I listen to NPR while chugging along in my sun-faded Subaru. But we both speak the language of rivers—a language older and deeper than political discourse. When we're standing together in the waters of the Colorado searching for stoneflies and trout, our opinions seem as fleeting as the crests and troughs of the waves that slip past us.

This book is not concerned with the unfairness of Bud being mislabeled. It is concerned with the dangers of dismissing people we enviros automatically assume are the enemy. A few years ago I wouldn't have spoken a civil word to an oilman much less spent time on a river with him. But I have come to believe this: An oilman who has a lifelong passion for healthy watersheds, and who combines this passion with the technical skillsets and financial means to preserve what he loves, can be a damaged river's best friend. Bud's experiences recovering oil and gas from the earth have trained his mind to identify problems and to solve them. The upper Colorado River is broken. Bud knows what broke it and understands how it can be fixed—and he is determined to see it fixed.

At RIM Companies, the oil and gas concern Bud cofounded, a large map of the upper Colorado River is pinned to the wall. Highlighted on the map in text boxes and diagrams are details of the wrongs done to the watershed. On a shelf near the map stands a statue of Don Quixote given to Bud by his father-in-law, who inscribed on its base, "You are now the family's Don Quixote."

A few years back, writing about an oilman tilting at windmills would have struck me as absurd. But water is the universal solvent because it dissolves so many substances; it can transform almost anything, even our most hardened beliefs. And rivers, given enough time, break through every barrier. When we recognize our common interest in rivers, the differences that separate us can dissolve.

Near the Colorado's source, a stonemason teams up with a telecommunications pioneer to initiate studies of the river's failing health. A grandmother undergoes chemotherapy while forging a river agreement for her rural county; her counterpart across the negotiating table is a water developer from Denver who worries about population's cancerous growth. And an oilman and an enviro wade together into the riffles and pools of an ailing river to find out what went wrong and to figure out how the damage can be undone.

Part I

Chapter 1

BOOKS OF NATURE

A fter draining the beer I was drinking at a bar in Boulder, I slammed down my mug and said, *"An oilman who wants to save a river?"* It wasn't the first time I had said this. Nor was this the first beer I had swilled while straining to understand an oil executive conservationist. I thought my head might explode.

Jason Hanson, scholar at the Center of the American West, beer aficionado, and one of the smartest people in the room when it comes to making sense of the New West, was unfazed. He gulped from his glass of amber suds and said to me, "I've never met an oil guy who doesn't love the outdoors."

As I spent time with Bud, I saw what Jason was getting at—and I realized that an oilman conservationist presents an intriguing paradox. Perhaps a hunger for extracting oil and a love of the natural world can be compatible. Maybe they can even be closely linked.

Many people in the oil business crave being outside, roaming the land far from office buildings. They start out as geologists because the lost worlds that lie buried beneath their feet enchant them. They speak of faulting and folding as if these events were plot twists in a gripping drama. They probe beneath the surface, deciphering clues stashed in stone strata and in the residues of ancient seas. They venture back through billions of years to Precambrian time when our planet formed; they explore the Paleozoic Era when life exploded on Earth; they journey through the Pleistocene, when the great glaciers advanced and retreated in cycles of shifting ice. They read pages from the stories of this watery sphere, and they understand that what is now desert once lay beneath the sea. They know the stability of the world is an illusion created by the shortness of our lives.

Jason reminded me that John McPhee, one of our favorite authors, had applied his art to the petroleum geologist's quest in his Pulitzer Prize–winning book *Annals of the Former World.* After we left the bar in Boulder, I started rereading McPhee's masterpiece and then tracked down *Oil Notes,* a book by Rick Bass, my favorite nature writer. Bass's descriptions of working as a young petroleum geologist eager to extract oil from the earth are as beautiful as any rumination on a free-flowing river he's written.

When I was a young man working as a rafting guide and backpacking through the wilds of the West, I announced to anyone who cared to listen how much I hated people in the oil industry. I thought they hated the natural world. But after meeting the people I assumed were enemies of the environment, I realized many of them had been hiking and fishing in the mountains longer than I had, and they loved being outdoors as much as I did.

Bud's appreciation of the outdoors has surfaced in every conversation we've had. Whatever topic we discuss, he circles back to his experiences with water that span the planet: from the swamps of Indonesia where he spent his childhood, to the jungles of Vietnam where he went to war as a young man, to the rivers of Alaska where he experienced transcendent beauty and mortal danger, to the upper Colorado watershed where he has watched the river slip toward death.

Bud chokes up when he talks about the American dipper, a bird he has watched disappear. Seeing a man who earned combat medals in Vietnam, and who has made a living as a petroleum engineer and an oil entrepreneur, get so tongue-tied with sadness at what has been lost along the river that he can hardly speak leaves a strong impression. Bud cares about the Colorado headwaters, and he understands that the disappearance of dippers signals an ecological crisis. But what of his business with oil and gas?

Before I met the rest of the cast of characters in the story—the stonemason who serves as a stalwart voice for the river, the grandmother who brokered a peace deal in the water wars across the Great Divide, or the biologist who sleuthed out the source of a mysterious disease—first I had to make sense of an oilman who loves a river. Or rather, I had to make sense of my own discomfort with people who drill for the dirty substance that fills my life with ease.

WHATEVER WE MAY THINK of the petroleum industry, we must concede it involves people providing something real—real fuel that powers real engines that do real work. Jerry Seinfeld, in a "Comedians in Cars Getting Coffee" episode with Howard Stern, says, "To me, America used to be a place that made steel and cars and had giant department stores. Now basically we produce amateur talent and people who judge amateur talent."

And the oil business isn't just about cars. Petroleum and natural gas power our universities and hospitals, our libraries and theaters. The Kepler spacecraft is orbiting the earth and aiming a lens at distant suns, discovering earthlike planets by the dozens. Some of those worlds are in the habitable zone of stars, meaning they could have liquid water that harbors life. These discoveries would not have been possible without fossil fuels.

And petroleum molecules are fabricated into everything from the clothes we wear to the materials we use to build our homes. If we got rid of everything manufactured from petroleum or transported with it, we would be left shivering in the cold without a coat, let alone an iPad.

Once we were a society that farmed and mined and felled trees. Now we celebrate start-ups that manage social media data. But when we look past the cyber smoke and digital mirrors, we still live in a world constructed from raw materials. For all our telecommuting, we still drive cars made of steel. For all the virtual worlds we create, we still live in structures built with concrete. And everything involves oil, from manufacturing the laptop I used to write this book to the delivery of pet food for my Labrador. Bud's business is responsible for impacting the environment by extracting oil and natural gas. But so am I each time I linger in a gas-heated shower.

The deeper I drilled into the dirty truth of oil and gas, the more I realized that demonizing the people who provide the stuff I demand is not the most productive way to channel my distress. To say I love a river, but I hate the oil companies that make it possible for me to drive to the river, is intellectually vacant. All of us are complicit in using fossil fuels to build a civilization that allows us to discover distant planets and drive to wild rivers and take hot showers when we get back home.

Some of our anger at oil companies is due to our displaced guilt at the gluttony of our fossil fuel consumption. And some of our hostility can be attributed to the unsettling reminder each time we use fossil fuels that we live in a real world of real consequences—we haven't severed ourselves from the physical earth. We haven't transcended the material realm into a cyber-world of electronic dreams that leave the planet as unsullied as the digital images we see on our screens.

The brute truth is we all use oil, and the filthy smoke of its combustion provides us with incalculable comfort. You can't heat your home with an app.

TAPPING THE CONCENTRATED energy in fossil fuels—billions of years of carbon-based life packed densely in chunks of coal and pockets of oil and gas—has allowed us to create a civilization hounded by guilt at the fuel foundation upon which it was built. In homes heated with natural gas we rail against fracking. We drive cars to attend oil protests. But my friends at the National Renewable Energy Laboratory in Golden, Colorado, who are among the smartest—and greenest—people I know, tell me they are still a long way from developing alternative energy sources sufficient to maintain our extravagant lifestyle and to continue our exponential increase in computing power.

The physical world is not as simple as the digital worlds we summon on our screens. For those screens to exist, now and in the foreseeable future, we will need to use oil and gas. And for oil and gas to fuel our world, real drills have to penetrate the real Earth. It's not a sexy story like windmills saving the world. It's a story of the world as it is. We can't just throw some solar panels on our roofs and go about our lives believing we have no impact on the planet. Photovoltaic panels must be manufactured and transported with fuel. And the toxic waste generated by building the panels has to be trucked away and stored somewhere—preferably out of sight and out of mind so we can believe the pristine power of the sun will liberate us from our reliance on the earth's resources and separate us from the repercussions of our greed.

But the sun doesn't shine every day. And sometimes the wind stops blowing when we need power most, forcing us to burn fossil fuels to fill these gaps. Building batteries to store clean energy is a dirty business, requiring mining toxic metals. We can cache energy by creating pumped storage hydroelectricity projects that move water to a higher reservoir when the sun is shining or the wind is blowing, and then release the water to a lower reservoir to spin turbines when power is needed—an elegant engineering solution. But this large-scale manipulation of our already stressed water resources wreaks havoc with the hydrology and biology of rivers and lakes.

Moreover, solar arrays and wind farms large enough to collect these weakly concentrated energy sources on a scale that
could power our cities would require enormous amounts of
land, as well as many thousands of miles of high-power trans-
mission lines, disrupting natural landscapes and disturbing
wildlife habitat on a vast scale. To keep the lights on, and to
keep consuming the iPads we use to post on social media how
much we hate fossil fuels, we would have to cover our nation
in a network of wind turbines, solar panels, batteries, pumped
hydro reservoirs, concrete platforms for wind turbines, service
roads, and power lines, leaving little room for escape into the
wilds, away from our clean energy infrastructure.

If we stop burning fossil fuels to slow climate change, and
we harness the power of the sun and wind so everyone can
have an iPad and a two-car garage, but the rivers are empty of
fish, and the forests barren of birds, what will we have gained?

INCREASING ENERGY EFFICIENCY, the mantra of many environmentalists, isn't the solution. It can even make the problem
worse. The more efficient our energy use becomes, the more
energy we use. Because jet travel is so much more fuel-effi-
cient than it was several decades ago, the cost of flying has
dropped—so now we fly more, increasing our overall consumption of jet fuel.[2]

Of course, we could all use less energy—this would reduce
our impact on the earth's climate and its finite resources. But
few of us seriously entertain this option. Who is willing to let
their laptop turn off when the sun goes down each day? Who
wants to live within their ecological means by walking everywhere—even to fracking protests?

When I lived in Boulder, considered one of the greenest cities in America, cars clogged the roads and packed the parking

2   The conundrum of increased energy efficiency leading to an
increase in fossil fuel consumption is documented in David Owen's
aptly titled book *The Conundrum*.

lots. While researching our energy future, I attended a conference concerned with preserving nature, in which speakers criticized oil companies, water utilities, government—they pointed fingers at everything but themselves. Many of the angriest people at the conference had eschewed urban density to live in exurban locales—arguably the least responsible ecological choice a human being can make, as it fragments the last scraps of wildlife habitat and requires long commutes. One of the surest ways to destroy nature is to live in the mountains with a view of nature. The most virtuous stewards of the earth's resources live in walkable cities.[3] A New Yorker who doesn't own a car has a lighter footprint than a fractivist who builds a house in the foothills above Boulder and drives to meetings to yell about oil and gas.

The parking lot outside the conference in Boulder, I am sad to report, was crowded with cars, and one of them was mine. I drove there solo because I was too lazy to arrange a carpool, or to fix the flat tire on my bike. None of us are innocent.

The more time I spent observing fracking protests, the more uncomfortable I became with people who wanted to ban drilling within sight of their homes. And the more I questioned my own intellectual honesty and moral courage. A few months earlier I had driven to those protests not as an observer but as a participant.

The protesters told me they used natural gas. It seemed they just didn't want gas wells within their narrow world. Had the wells been drilled in the Piceance Basin of Colorado and threatened pristine rivers in that distant place, they would have stayed home. Nor would they protest wells drilled in Nigeria, or any other place on the planet where the curse of oil in developing countries leads to apocalyptic environmental conditions and human rights abuses by people in power who pocket the oil and gas wealth—and the people at the bottom get less than nothing in return. Their children are left breathing toxic fumes and drinking contaminated water as their poverty deepens.

---

3   This idea is explored in convincing detail in *Green Metropolis* by David Owen.

And if they complain, the repercussions can be horrific.[4]

When oil and gas—which every one of us in America uses every day—is extracted from our backyards, we are protected by some of the strongest environmental laws in the world. And the wealth that's generated is distributed among people who own mineral rights, and is spread through taxes that support local economies—not concentrated in the hands of corrupt governments that abuse their citizens.

As a citizen of the world, I am not comfortable telling a Nigerian mother the fuel I use should be drilled in her backyard instead of my own. My children will be much better protected from the consequences of extracting oil and gas than her children will be. Expecting her to bear the burden of my energy use is morally outrageous.

Nor am I comfortable telling peasants in China who live near hissing lakes of toxic chemicals—created by processing rare earth metals needed for the magnets that run motors in wind turbines—that their health is less important than my own.

I prefer the oil and gas I consume not be drilled in a distant country where I will never see the suffering of people brutalized by my energy use. Until the day I stop using oil and gas, please drill in my backyard. And process the dangerous materials needed to capture clean energy in my backyard. Not NIMBY but YIMBY (Yes In My Backyard). If the energy development that allows my world to run is kept in my sight and in my mind, I will be motivated to reduce my energy use, instead of paying lip service to the principle. And each day that I hear the roar of the wells and inhale their fumes, I will be reminded of the great tragedy of our fossil fuel consumption: not that we have burned oil and gas to launch satellites and to build MRI scanners, to heat schools and to power hospitals, but that each one of us has squandered the earth's bounty so we can drive from overbuilt homes to overstocked stores to purchase trinkets and gadgets we throw in the trash.

---

4   This curse of oil development in poor countries is documented in *Crude World* by Peter Maass.

WHEN I THINK OF ENERGY companies, Exxon and BP come to mind. It's easy to hate these behemoths with their bloated profits and cleverly orchestrated propaganda campaigns. But more than 70 percent of domestic onshore oil and gas production is from independent companies, not from energy giants with their logos on gas stations.

Small farms and local craft producers have been celebrated by sophisticated consumers, but small-scale energy-production companies are not what trendsetters who tout "buy local" have in mind. Praising a locally grown tomato is popular— and so is denigrating the locally sourced oil that lets us drive to farmers' markets. Artisanal energy crafted from fossilized sunlight is not likely to capture the public's imagination.

And speaking of capturing the public's imagination: Josh Fox of *GasLand* fame would have us believe we can build a bunch of windmills and everything will be fine. That's a lovely story. Tell it to birders who would witness the avian slaughter. Tell it to lovers of bats, which would die in droves. Explain it to folks fighting Pebble Mine in Alaska, which would damage one of the planet's last great salmon fisheries to source copper needed for wind farms.

Every effort to power our world, no matter how well-intentioned, has unclean—and often unforeseen—consequences. We are as likely to develop an energy source that has zero impact on the environment as we are to build a perpetual motion machine.

"Solar roads" are all the rage these days. But when the initial euphoria of the idea wears off, we'll laugh at the silliness of putting solar panels in roads, of all places. A bright grade school student understands that placing solar panels *above* roads to capture sunlight makes more sense than *driving* on the panels, covering them with road grit to block the light and scratch the glass. But solar roads could power the entire

United States, their proponents insist. How the energy generated by the solar roads would be stored and transmitted no one has explained—solar roads are too cool a concept to agonize over technical details. Such are the discussions about our energy future: We chase implausible dreams and hurl vitriol at the people who give us oil and gas. Like so many others, I have let myself get caught up in solar road hysteria. I dream of a clean energy future, but tainted reality intrudes.

After researching the real world of renewables, I would like to challenge Josh Fox to (A) stop working in a natural gas–heated office, to (B) spend some time living with the parents of children in developing countries devastated by serving as our energy colonies, and to (C) make a film about the downsides of renewable energy so we can have a realistic discussion about how to move forward as a civilization that values unlimited consumption more than the preservation of the planet.

Josh Fox's entertaining road trips around the country to document the dangers of energy production, and the natural gas he consumes to heat his office, are as much a part of the problem as Bud's company drilling for the oil and gas that power Mr. Fox's world—which runs on equal parts self-righteousness and fossil fuels. Complaining about oil and gas has become a cottage industry, and a lucrative one for people like Josh Fox. Their loud complaints, like climate change denial on Fox News, are sound and fury with little substance. They move us no closer to reasoned discourse and substantive solutions.

Not to single out *GasLand:* The hypocrisy at the heart of the American environmental movement evolved several decades ago. Edward Abbey, the radical saint revered by enviros, paraded his elephantine Cadillac through the streets of the Southwest as he penned prose that encouraged activists to monkeywrench the infrastructure of industrial civilization. Abbey's penchant for gas-guzzling cars was the moral equivalent of a preacher who pounds the pulpit about chaste living getting caught with hookers. And it epitomizes the childishness tucked deep in the hearts of many American enviros, a mindset of toddlers: Give me my toys, and remove anything I don't like from my sight.

I stashed a dog-eared copy of *The Monkey Wrench Gang* in my waterproof ammo can on rafting trips I guided on the White Salmon River in Washington, when I was young and angry and wanted to blow up dams. Now the film *DamNation,* which I contributed to as a consultant, features the demolition of Condit Dam on the White Salmon and is inspiring young activists to take up the cause of tearing down dams to restore rivers. But the "deadbeat dams" that the film derides—dams that have outlived their usefulness and now provide little economic incentive to keep them standing—are relatively few. "Damolition" is a limited tool that no sane society would use to destroy carbon-neutral power producers like Glen Canyon Dam.

However, climate change and overuse of the Colorado River could drain Lake Powell behind the dam to "dead pool"—the level at which the turbines can no longer generate power to keep air conditioners humming in Phoenix.

The most important energy-related film or book that could be produced right now would encourage young people to become scientists and engineers and entrepreneurs working to solve our energy and water challenges: something that recalls America's get-it-done-ness before professional complaining replaced industries based on problem-solving. I've done my fair share of complaining about people like Bud. And my protests have done precisely nothing to make the world a better place.

My love for rivers has remained constant. My understanding of the methods that will save them in the real world has changed. Edward Abbey driving his gas-guzzling Cadillac and Josh Fox sermonizing in his gas-heated office offer us entertainment, not solutions. And one thing we could do without more of in America is entertainment.

When gas is released as a byproduct of coal mining, methane becomes a pollutant that accelerates climate change. A company Bud invests in captures the waste methane to use as a source of energy—the gas turns turbines, generating electricity that Aspen uses to run ski lifts. This isn't a perfect energy solution. But it's a good start. And making perfect the enemy of good gets us nowhere.

Bud says he looks forward to a future in which some advancement like cold fusion disrupts his industry and replaces fossil fuels. But he argues that a lot of energy is still stashed in the earth in such a pure and portable form, we would be smart to use it in the cleanest way possible to power our advancing technology as scientists, engineers, and entrepreneurs continue to develop substitutes for the fossil-fuel diet of our energy-hungry civilization.

Up the road from us at Colorado State University in Fort Collins, a chemistry professor, Amy Prieto, has launched a start-up company to create a new generation of batteries that will last longer and not require toxic acid to produce. Developments like this could revolutionize the way we use energy and reduce our impacts on climate change. But these breakthroughs take time—and scientists solving problems make for dull films. Transforming our energy infrastructure to accommodate new technologies, while still providing the standard of living we demand, is more complicated than stories like *Gas-Land* would have us believe.

Most energy experts, no matter how green, concede that the drilling platforms of the oil and gas industry are not going away anytime soon. The slippery blackness of oil will continue to flow through our lives for many years to come, no matter how aggressively we pursue alternative ways to power our world, no matter how much we want magical solar roads to save us from ourselves.

THERE IS NO QUESTION THE climate is changing—it always does. Tree rings help researchers reconstruct the West's climate before scientists arrived with instruments to measure weather and river flows. Written in the records of trees is a disturbing truth: The West was much drier before we settled the region. Decades-long droughts scorched the West in the distant past. We arrived on the scene at an unusually wet period and assumed it was the norm. The Ancestral Puebloans, who left no written records in their desert cliff houses of the Southwest, may have thought the same thing. The abandonment of their stone cities at the end of the thirteenth century coincided with a drought that dragged on for decades.

What we thought was an arid climate in the West as wagon wheels cracked the crust of dry earth was, in the long view, relatively lush. The drier average across the ages will almost certainly return to the region—and may be doing so right now as reservoirs on the Colorado River steam to nothing in desert air and snowpack shrinks across the region.

Jim Yust, a fourth-generation rancher with a spread on the upper Colorado, says when he was a boy, the family's barbwire fences in winter were always buried past the fourth wire. Now, the snowpack seldom reaches the third wire. And the anecdotal evidence of old-timers is supported by the models and data of scientists, who warn that the West has been heating up in recent decades and will get much hotter. As temperatures rise, evaporation will increase and soil will dry out, reducing the amount of runoff in rivers, shrinking their flows. And in the hotter years ahead, the snowpack will be smaller—this is bad news not only for the ski industry. Snowpack thickens throughout the winter as blizzards blanket the Rockies; when the spring sun shines on the mountains, melted snow pours down from peaks to the plains and valleys below. Snowpack is a reservoir, and many forms of life depend on its fullness.

Even if precipitation totals in the West remain the same or rise as the climate heats up, rain, rather than snow, will fall. Water stored in snowpack is held in reserve for many months; water from a rainstorm soon disappears downstream. Snowpack is already melting earlier in spring—this crucial reservoir is shrinking as the world warms. As a consequence, the Colorado River is suffering.

There is little doubt Bud's work contributes to climate change, which is diminishing the mountain snowpack at the Colorado's source. There is also no question I contribute to climate change whenever I turn the key in my car's ignition, each time I order cheese nachos, every time I purchase a new pair of running shoes. Countless small choices I make each day reset the planet's thermostat as dramatically as a drill that pierces layers of rock to release the carbon-rich remains of vanished worlds.

Saving a river is a futile task, for this planet will never stop changing—regardless of whether we curb our greenhouse gas emissions. When you squeeze the sand of the Colorado River Delta, the remains of the Grand Canyon trickle between your fingers. Time and water have done this work. All that we love will be folded back into the earth and return as something else. This doesn't stop Bud from trying to restore the river to some semblance of what it once was. As an earth scientist he understands he cannot save the river forever. But he can work to make sure his grandchildren have the chance to pinch a caddisfly cocoon between their fingers and watch dippers dive into the water to hunt insects. He wants his grandchildren to have the opportunity to read these messages nature has taken millennia to write.

Bud speaks of leviathan trout making wakes as they turned to give chase to stoneflies and drakes. "Notice that I'm talking about all of this in the past tense," he says, shaking his head and running a hand through his hair. He still has the strong jawline and solid shoulders of the football player he was as a young man. But whitish hair covers his head in thinning wisps, and the plastic cable of a hearing aid wraps around his ear. He tells me that when I understand what this river once was, and when I grasp what it has become, then the gray in my own hair, now confined above my ears, will spread to cover my head.

In the days before the upper Colorado succumbed to sickness, Bud would crouch at its edge with a caught fish while trout shot from the water like missiles streamlined and bright. He'd reach into his net to cradle the plump belly of a fish, tracing the crimson stripe of a rainbow, or the haloed dots of a brown trout, committing the details of these creatures to memory—and these memories he would someday paint, fixing the images in pigment and paper to stop their beauty from fading.

AUTHOR HARPER LEE, ALMOST a half-century after the publication of her novel *To Kill a Mockingbird,* wrote this in a letter to Oprah Winfrey: "In an abundant society where people have laptops, cell phones, iPods, and minds like empty rooms, I still plod along with books. Instant information is not for me. I prefer to search library stacks because when I work to learn something, I remember it."

"Rivers are like books of nature," says Bud. He points out that lessons written in their waters cannot be learned on screens. And similar to how our reliance on the virtual world diminishes our appreciation for extracting energy from the earth and manufacturing things, we tend to forget that the natural world is real, and rivers run through it.

Bud says, "People today prefer science fiction and virtual craziness to the real world." He worries our society is becoming so narcissistic and addicted to self-promotion people won't see beyond themselves to appreciate and preserve natural resources. He worries about young people becoming disconnected from what is real—the physical world of oil and steel, the natural world of rivers and wildlife. He worries children will lose themselves in pixilated screens and will never know the feel of wind that wrinkles the river pushing past their cheeks, will never hear cottonwood leaves crackle beneath their boots and smell their sweet decay, will never see the sharp wake of a trout as it turns to give chase to its prey.

Bud's lifelong obsession with water, and his determination to make sure the next generation inherits a healthy Colorado River, became clear to me when I drilled down into his early life, spent largely in oil camps.

Chapter 2
OIL AND WATER

When Bud's father moved the family to an oilfield in the rain shadow of California's Diablo Range, Bud chased horned toads through the hot sand and listened to railroad cars rattle through the desert. Some of those train cars carried potable water to the remote outpost where the family lived in temperatures that soared to 120 degrees. Beneath the parched land lay oil—liquid remains of ancient life poised to fountain skyward when engineers sank their wells.

In the oil camps of Bud's childhood, his curiosity was given room to roam, and from his restless mind erupted geysers of inquisitiveness. When he wondered if cotton from cottonwood trees would burn, he lit a handful on fire—answering his question and burning down a barn in the process. Bud had no television. He entertained himself by building CO_2-powered balsa racecars and plinking with a .22 caliber rifle given to him by his grandfather on his ninth birthday—about the time Bud began his lifelong love of riding horses. At age seventy-two, when he began telling me about his struggle to restore the upper Colorado River, he was recovering from twelve broken ribs after being bucked off a horse.

When Bud's father was transferred to Indonesia, the family boarded a boat and traded the baking heat of the San Joaquin Valley for the dripping humidity of Java. Bud was immersed in a water-filled world unlike California's sere landscape. Among his memories of the long boat journey to Java are waves in a typhoon that towered to four stories and sampans clustered in the Hong Kong Harbor to form a floating city. Divers speared their slender bodies into the water to retrieve shining coins tossed by travelers, and workers collected human excrement to fertilize their crops. At age eleven, Bud studied each detail as he tried to make sense of scenes for which television had not prepared him.

In Indonesia, Bud faced formative experiences as diverse as being caned for making mistakes with spelling in the British school he attended, seeing lepers with noses sheared from their faces begging in the streets, and sitting down to dinner in the family's home with visitors that included ship captains and the novelist James Michener.

In the gritty city of Jakarta, Bud's family lived in a house abandoned by the Dutch, who had been purged when Indonesia gained independence from its colonial rulers. The toilets in the family's home flushed into the city's open sewage trenches. Few things were more entertaining to the scatological mind of a little boy than watching a turd disappear down the toilet and then reappear in a ditch outside to sail away like a foul boat amid the city's fetid waters. People in the streets squatted over open canals to relieve themselves, while downstream, workers washed laundry.

Amoebic dysentery was rampant; securing a safe water supply was a constant concern. The region's abundant rains were gathered in a cistern on the family's roof. Bud watched mosquito larvae twist in the water, which was boiled and treated with iodine. Bud's lifelong fascination with insects has served him well as a fly fisherman, and as an advocate for the Colorado River.

The family moved from the bustling city of Jakarta to an oil camp deep in the jungles of Sumatra, where deadfall was taller than an elephant's eye and prowled by tigers with silent paws. Bud busied himself by studying the tracks of tapirs, listening for the crashing approach of an Indian rhinoceros, and catching Siamese fighting fish in a stream that fed a reservoir. The swampy water around the camp was stained with tannins the color of cola. With a machete in hand, Bud delved into the jungle each day, his boots squishing in the soggy clay as he studied leafcutter ants and lizards as long as his arms. Howler monkeys screamed the family awake each morning, and parakeets and parrots flew past their house. A cobra slithering across an office floor sent everyone scurrying outside, and locals skinning a twenty-nine-foot python lies coiled in Bud's memory. He made a pet of a mongoose, a sharp-toothed creature that could hold its own against deadly snakes.

Bud drove with his father in a jeep deep into the jungle at night, looking for glowing eyes in the beam of their light. In daylight they hunted wild boars. Bud says he has always been a hunter, whether pursuing game in a Sumatran rainforest, reading a river's flow to find fish, or hunting deals to build his business. The pursuit of oil is seldom a success: Even the best at it miss more often than they hit.

Together father and son went into the field on engineering tasks. Bud got hooked on the adrenaline rush of finding oil when he watched wells abandoned by the Japanese in World War II being reopened. Some wells gushed oil and gas a hundred feet into the air. To find these fluids and release them from dark cellars was great adventure for a boy. Bud grew fascinated with pulling liquid from the earth. He says petroleum engineering is basically pipefitting—a form of plumbing. His formative years were defined by liquid, by its absence and its abundance, its power and its peril—and he is a lifelong plumber. He has been known to sweat a pipefitting in his home for fun.

In the Sumatran oil camp, Bud attended a Calvert school with one teacher who covered kindergarten through eighth grade. Whether shaped by the Calvert system, which aims to foster a lifelong love of learning, or by his childhood abroad, Bud's drive to figure out how things work was essential in the second half of his life, when he set out to understand how the Colorado River was being stripped of its rich array of creatures. His boyhood lessons teemed with flora and fauna that led to no end of fascination. In jungles where standing pools stank of rich life and wetness jeweled the leaves, Bud became a student of the earth, of its creatures and its waters.

Each day in Indonesia Bud was surrounded by people determined to figure things out—from his mother, a fastidious woman of Germanic lineage who ensured the family had safe water to drink, to his father, who recovered oil from the concealed history of the earth. Oil is always trying to climb toward the surface, to rise to the place of lowest pressure, but is blocked in its upward flow by roofs of impermeable stone. Petroleum engineers figure out how to puncture this caprock, freeing the oil to migrate where it wants to go—upward toward the light.

Petroleum engineering is a science. It is also a treasure hunt that relies on intuition—a quest that rests on hunches to probe places where oil has hidden for hundreds of millions of years, and to pry open these unseen vaults. Into the bright day are drawn the sludge-black residues of creatures that slithered and crawled in the Cretaceous, the slippery remains of algae, the transformed remnants of ferns that once shivered their fronds in the wind. Oil tells tales of long-ago worlds, when islands sank beneath oceans and seashores lapped at desert dunes. Gas rushes upward from wells, cracking the air with a jet engine roar as atoms of hydrogen and carbon, once linked to form life, are released. An earth scientist gains the insight of a mystic: At the most basic level, all things are connected.

Extracting oil is a business. It is also a pursuit that stirs the playground of an engineer's mind. Punching pipe through the earth to coax oil and gas up from the dark is an exhilarating game. Bud's imagination swooped through eons of geologic time, exploring fault lines and tunneling through pores beneath unthinkable pressure. Oil floats on water, and these fluids are usually found together. In a dome buried deep below, oil rises to the crest, hovering atop water from ancient seas.

Colorado School of Mines molded the mind of a curious kid into the brain of a trained engineer. The coursework was brutal for Bud, who struggles with dyslexia. He was always on the edge of flunking out. But he persevered and graduated with a degree in petroleum engineering. He jokes that Mines (as this MIT of the West is known) was so tough he would rather go back to Vietnam than return to a classroom at Mines.

Bud was in the Reserve Officers' Training Corps at Mines; when the war in Vietnam began, he volunteered to go. At age twenty-five he found himself in charge of a platoon of thirty engineers in the jungles of Vietnam, where humidity pearled on leaves as large as satellite saucers, and the puddled mud bred swarms of mosquitos. One of his unit's tasks was providing a supply of safe drinking water. An army marches on its stomach, Napoleon noted. But even with plenty to eat, a soldier without water will die.

Bud was the first non–West Point officer to be assigned combat duty with the 101st Airborne Division, a position sought after by "ring knockers," graduates of military colleges and universities. He pulled this off by dating a general's daughter—and he married her after returning from his service in Vietnam, where he earned a Combat Infantryman Badge, Silver Star, Bronze Star, and Purple Heart. A fight with the Viet Cong Bud led is featured in the book *Battles in the Monsoon* by military historian S. L. A. Marshall.

Amid the chaos of war, killing enemies is easier than capturing them. During Bud's battle, his troops became the first group of American soldiers to capture a significant number of North Vietnamese prisoners. In the crucible of Vietnam, Bud discovered in himself a willingness to accept risk if he believed what he was doing was right. "I realized I was willing to stick my neck out," he says. He went to war because he believed his country was threatened by the spread of Chinese Communism throughout Southeast Asia, and he felt a connection to that part of the world after spending so much of his childhood in Indonesia. He took prisoners because it was the right thing to do—even though it was the hard thing to do.

Bud traces his fight to preserve the Colorado back to battles in Vietnam. And as a river can be followed in narrowing streams toward its headwaters, Bud's determination to save the source of this endangered river can be traced even further back, to a storm in Louisiana.

Before he went to Vietnam, Bud worked on an offshore oil rig. He stayed on the platform for twenty-four straight days, until being awarded his own rig to run, but was forced to abandon it when a hurricane hit. "I made sure I was the last man to leave my rig and the first one back on," he says. After the hurricane passed, he had to swing on a rope from a crew boat to get onto the platform and assess the damage. The rig was his charge; he was responsible for protecting it and making sure it did the job it was designed to do.

When Bud had been a little boy in jungles thick with danger, his parents never doted. "They let me look out for myself," Bud says. "They wanted me to learn to protect the things I cared about." When the Colorado first flowed through his life, he developed a bond with the natural systems supported by its waters. "I had to take care of the river," he says. "It was like the oil rig. I had to protect it." When he realized the river was not functioning properly, his instinct was to restore its systems to working order. A lifetime of watching the movement of water and studying the natural world led him to a simple fix: building a bypass around the barrier that blocked the river's flow.

But when it comes to water in the West, sensible solutions are rarely implemented without long and bruising battles.

⁓⁓⁓⁓⁓⁓

WHEN BUD'S FAMILY RETURNED from Asia, he listened diligently to an uncle who was Dean of Engineering at City College of San Francisco. He told Bud he believed the biggest limiting advance to civilization would be the world's finite amount of freshwater. He predicted California was on the cusp of water wars as dry Southern California experienced unsustainable sprawl and made moves to tap the liquid bounty of sparsely populated Northern California. His logical mind reeled at the shakiness of building a civilization where a resource human beings cannot live without is in short supply. He predicted no end to the problems that would ensue.

Not until nearly thirty years later would Marc Reisner publish *Cadillac Desert,* the landmark book that opened the nation's eyes to the poor planning that led to building civilization in some of the driest places on the planet. What is now accepted as painful truth—that we have made a mess of our scarce water resources in the West—was wildly prescient when uttered by Bud's uncle in the 1950s. Dams of gargantuan scale were being built at breakneck pace, and few Cassandras predicted the water crisis the West would face. Bud's uncle spoke of Northern California breaking away from the state to protect its water from the thirsty hordes of Los Angeles and its parched surrounds.

The strangeness of packing people into places without adequate water supply, after reaching its illogical extreme in California, replayed itself in the state of Colorado—where the water flows on the west side of the Continental Divide, but the people live on the east side. Colorado's mismatch between population and water supply would drive the conflict that prompted Bud to battle over the headwaters of the arid West's most iconic river. But as a sophomore in high school, he couldn't foresee any of this, of course. He just listened as his uncle spoke of the West's coming water wars. And this warning from one of the wisest men Bud knew merged with lessons from a water-focused childhood that spanned the planet.

"Oil and water are closely related," says Bud. "One creates energy, one creates life." He is convinced that as water resources around the world face increasing pressure, "people will realize that water is more precious than oil." Recently he was appointed to the Board of Governors of his alma mater, Colorado School of Mines—a university at the forefront of developing energy-efficient water treatment technologies that make the best use of limited supplies.

From trains delivering drinking water in the oil camps of a California desert to soldiers in Vietnam bathing themselves with jungle water collected in helmets, Bud's experiences with our most vital fluid taught him this resource should never be taken for granted. Rivers run full and free, until one day they don't. And then we are left staring at a dead ditch where once a living river flowed, wondering what went wrong—and worrying about our own demise. Bud is a person trying to save a watershed. He is also part of a watershed trying to save itself.

Chapter 3

WHAT ONCE WAS

John and Ida Sheriff are two of the last people alive who can share what Grand County, Colorado, was like before water diversions to the Front Range began drying up this scenic valley ringed by snowy peaks. In the spring floods of their childhood, boats sailed across meadows. When the snowpack melted in spring and rushed down from the mountains, the water refroze, creating ice jams, John explains. He hooks his thumbs in his overall pockets and slumps low in his chair as he reminisces. Water would back up behind the frozen barriers, saturating hay fields. Then these ice dams would melt, spilling blocked water back into streams that feed the Colorado. Bloated and brown, the river would crash across its banks, twirling downed trees like toothpicks. Spring runoff now is a trickle compared to the torrents of the past, when the Colorado River was called the Grand River and gave its name to Grand Lake, a waterbody near the river's source.

Grand Lake was known as Spirit Lake to the Ute, who believed the souls of slaughtered enemies dwelled in its depths and rose through mist that swirled across its surface. The waters were shockingly cold and as transparent as air. Long after the Ute were pushed out of the valley, the lake was sullied by the construction of an adjoining reservoir. Filled with weeds and silt, this artificial basin spilled its tainted water into the clearness of a natural lake once at the center of the Ute's spiritual beliefs. The Ute had displaced other tribes from their summer hunting grounds, and were themselves chased away from this abundance. Pushed farther and farther west, away from lush mountain valleys, the Ute were forced to settle in desert lands deprived of water.

The Grand River also gave its name to the county that cups the Colorado headwaters in a valley perched between plains and sky. Grand County is located in Middle Park, named by French-Canadian mountain men who called a mountain valley a *parc*. Into this grassland rimmed by mountains and rich with game they came to trap beaver.

Similar to the bison that once blackened the plains in their multitudes but were decimated in a few decades, the beaver colonized every reach of river in the Rockies—and all but a smattering of them were killed in a handful of years as their soft pelts became hats to feed a fashion craze. The infrastructure that these industrious creatures assembled by felling aspen trees for use as construction material once held back the floods of spring. Mountain snowmelt swelled behind the waterworks of beaver and seeped into meadows to nourish stands of aspen harvested by these hardworking animals.

We have only the faintest hints of the hydrology and ecology of the West that evolved through the millennia, for the mountain men eliminated the beaver in an evolutionary eyeblink. Their dams toppled. The bloated rivers of spring surged down from mountains, damp meadows crusted with dryness, and animals nourished by wetlands fled to moister places beyond the reach of men. Chains of life shattered when the steel jaws of traps snapped closed on beaver. Without this creature's waterworks storing spring runoff, the water cycle of the Rocky Mountains was retimed to a new rhythm.

After the West's original dam builders disappeared, explorers arrived. They made maps and documented the region's flora and fauna—which had been altered long before they followed paths tamped by the feet of a species even more industrious than beaver. Humans had hunted mammoths, giant sloths, and the continent's other megafauna to extinction with spear points thousands of years before the mountain men slaughtered beaver with their traps. When pioneers arrived in Middle Park, what seemed like wilderness was as much a product of human nature as it was a creation of "nature." And when a new breed of dam builder in the twentieth century replaced the sticks of rodents with concrete walls, the pace of reengineering the region accelerated.

JOHN SHERIFF'S GREAT-GRANDMOTHER quit the mining boomtown of Leadville, Colorado, to make a go of homesteading in Grand County in 1881. On land near Hot Sulphur Springs she moved into an abandoned cabin and had three sons. One of the boys was John's grandfather; John's father was also born on the property. Homesteaders in this snowy place that soars nearly 8,000 feet above the sea raised some food for subsistence but survived mostly on supplies sent up by freight line from the friendlier climate of the Front Range. The Sheriff family grew hay and raised cattle, which they sent from Middle Park for sale in the cities of the plains. John was born and raised in a ranch house on the homestead, and only once did he leave the Sheriff property for a significant amount of time—when he joined the U.S. Navy and served in the South Pacific during World War II.

John met Ida when he returned from the war. Their parents had known each other for many years, but John and Ida first talked when they both put on their best clothes and attended a dance at a bar in town. They've been married and living on the land homesteaded by the Sheriff family now for more than sixty years.

Ida's parents also homesteaded in Grand County near Hot Sulphur Springs. Of Austrian ancestry, they felt at home surrounded by alpine peaks when they arrived in 1912. They came with the last wave of homesteaders, explains Ida. She is a diligent keeper of Grand County history; her bright eyes sparkle when she spins tales from the valley's past. The town of Hot Sulphur Springs was founded in 1860 on tourism, she explains. William Byers, a newspaperman from Denver, decided hot springs in a mountain paradise would be a big tourist draw. After "discovering" the springs tucked behind peaks that rise northwest of Denver, Byers duped Native Americans, who used the mineral-rich waters for healing, into selling the town site in a shady deal. And tourists from Denver soon displaced the Ute tribe.

Riding a steam train that smoked its way around switchback loops in enchanted mountains to soak in the sulfurous waters of a snowbound valley held allure for intrepid travelers. Winter carnivals in Hot Sulphur Springs featured feats of ski jumping—the first stirrings of an industry that would one day transform the economy and culture of Colorado's mountain towns. But the reality of riding in a train that chugged toward clouds on Rollins Pass at 11,660 feet, climbing through tilted terrain blasted with blizzards that buried telephone poles to their tops and sent avalanches crashing across the tracks, proved too daunting for most Denverites.

Not until the Moffat Tunnel pierced the Continental Divide in 1928, allowing trains to pass swiftly and safely into Middle Park, did tourism become a powerful engine driving the local economy. The storied winter carnivals of Hot Sulphur Springs, in which John competed as a ski jumper, stopped when the Winter Park ski resort, serviced by the train, started running lifts up mountainsides so tourists could schuss down the slopes.

The Great Depression of the 1930s stalled Grand County's tourism industry. Rugged folks who'd homesteaded the high mountain valley had to revert to survival mode. They were used to being poor, so it wasn't a big change, explains John. A crop of cabbage kept the Sheriff family from starving. John says, "My parents didn't have much but they always fed hungry people in those hard times."

John began working as soon as he was old enough to lift a shovel and tag along with his dad. As an adult he rode daily on horseback across open range to look after the family's cattle. He had 17,000 acres to cover and needed to move the herd constantly between water and grass—Ida didn't see him for weeks at a time. Not until the late 1950s did the horsepower of machinery replace real horses for cutting hay on the ranch. Nearing his tenth decade, John still cuts the same fields of hay.[5]

Ranching in these high altitudes is not for the faint of heart. Although a tough way to make a living, ranch life has for more than a century attracted scores of dabblers who play at riding horses and rounding up cattle. Tourists have always been the most lucrative crop to cultivate in Grand County, which has one of the shortest growing seasons in the lower forty-eight states. The stretch of scenic valley between Granby and Grand Lake became the dude ranch capital of Colorado. After a road was paved through Rocky Mountain National Park in the 1930s, well-heeled wrangler-wannabes arrived in motorcars. An airstrip where Charles Lindbergh landed when visiting the financiers of his historic flights at their ranch in Grand County now lies buried beneath a reservoir.

The history of Grand County is a story told across the West. Mountain men slaughtered beaver, pathfinders followed in their footsteps, prospectors swung picks in search of precious metals, native people were pushed to marginal lands, loggers chopped down forests, and ranchers eked out a living in brutal weather. Now CEOs with SUVs trade stocks on their smartphones as herds of Hereford graze postcard pastures and kids take a year off college to play in powder stashes and race mountain bikes down trails.

"There are no real ranchers left in Grand County," John says, shaking his head. Rich hobby-ranchers use the money-losing business of raising cattle as a tax write-off and a chance to dress like a cowboy. John and Ida gave up on cows when the Bureau of Land Management forced them to sell a right-of-way that became a road. Miscreants used the road to

---

5   John died at age eighty-nine in April 2015.

harass cattle, chasing the spooked animals in jeeps and with dogs, sometimes shooting them for fun. Life has never been easy in Grand County. But Ida says, "We feel blessed to be surrounded by such beauty—and by so many stories." Newspaper insulates the walls of old cabins on the Sheriff property. The writing on those walls now serves as a rich repository of lore from Grand County's past.

There was a time, Ida tells me, when lettuce grown in Grand County was served on plates at the Waldorf Astoria in New York City, so famous were the tasty greens that thrived in the Colorado sun and cool mountain air. But cheap lettuce grown by agribusiness ventures in California's Central Valley killed Colorado's commercial lettuce trade. Ida's sweet voice takes on a sharp edge when she speaks of transplants to Grand County from California and elsewhere who believe blight put an end to local lettuce farming in the 1950s. She tries to educate them, but these newcomers "believe what they want to believe." Ida nods toward John and tells me, "When you get to be our age, you'll realize that people change history all the time."

Wallace Stegner, one of the West's most noted historians and writers, when asked what a newcomer to California should know, answered in four words: "Water. It's about water." He might as well have added: "And infrastructure financed by the federal government."

Industrial farms in California's Central Valley became so proficient at growing cheap lettuce and a cornucopia of other vegetables, fruits, and nuts—about half the total consumed each year in the United States—because of a water project funded by U.S. taxpayers. The Central Valley Project, a complex of dams and canals built beginning in the late 1930s, was conceived by California to put a stop to water shortages and devastating floods. When the Great Depression emptied the state's coffers, the feds stepped in and footed the bill for a scheme of immense proportions—setting the stage for federally financed megaprojects that transformed the waterscape and economy of the American West.

The driving force behind California's Central Valley Project, one of the largest public works projects in human history, was the settlement of arid lands by small family farms. But when the project was completed, subsidized water flowed to sprawling corporate farms with absentee landowners. And when crates of leafy greens watered with taxpayer dollars shipped across the nation from California, the children of homesteaders in Grand County, Colorado, who cultivated lettuce to support their families, had to find other ways to put food on the table.

The law of unintended consequences when we intervene with good intentions applies not only to economics but also to ecology. When we alter the natural order of things, there are always repercussions—some expected, many not.

Windy Gap Dam, completed in 1985, turned a meadow along the Colorado's banks into a muddy swamp. Ida reminisces about the pest-free days before mosquitoes bred in the muck of the blocked river. A campaign to knock down mosquitoes sprayed pesticides in breeding grounds formed by the dam. Now concern is rising over the pesticides killing not only mosquitoes but also bugs that benefit the river by providing food for trout.

The good news about swampy Windy Gap Reservoir is it created perfect habitat for pelicans and cormorants. The bad news is pelicans and cormorants feast on trout in the river. Managing a watershed is an endless game of whack-a-mole: Each time you knock down a problem, another one pops up.

"We've done a lot of good things as a species," says Bud. "And we've made a real mess of a lot of things, too."

As Bud and I get up from the kitchen table at John and Ida's house to leave, Bud says to Ida, "My legacy before I go out of this world is to have the river flowing free again."

Ida shakes her head and gives a small laugh. "Oh, Bud, *good luck* with that."

"She'll only believe it when she sees it," Bud says to me. "And I don't blame her a bit."

Ida and John have witnessed water controversies along the upper Colorado River for as long as they can remember.

They have watched creeks that once ran full turn dry as desert roads. They have seen pumps move water into irrigation ditches once filled by gravity before water was diverted to the other side of the mountains. John's father was a Grand County Commissioner; he fought to protect the water rights of local people from being infringed by Front Range projects.

"Water is the beginning of a lot of wars," says Ida.

From neighbors skirmishing over irrigation ditches to residents around Grand Lake complaining that the once-clear water has been sullied by diversions, Ida and John have lived through almost every type of water conflict conceivable. Bud's battle to reverse the river's failing health and save the trout and other creatures that rely on its flow is a new chapter of an old story.

John and Ida were always too busy working on the ranch to catch fish in the river that runs a stone's throw from their front door—a reach once considered one of the finest rainbow trout fisheries in the West. But in the never-ending struggle to make a living in this gorgeous but stingy land, John's family started renting out cabins on the river's shore in 1932. For a dollar a day, fishermen could sleep in a rustic shelter crawling with mice and cast for trout as long as their arms. Those dollars helped the Sheriffs keep their ranch.

Ida and John remember feasting at community fish fries supplied with suckers and trout pulled from creeks in seine nets that grew so heavy with fish they had to be hauled by horses. But what brings tears to Ida's eyes is not the valley's vanished abundance of fish but the transformation of a nearby meadow.

A few miles upstream of the Sheriff ranch, the sculpted hills of a golf course rise from a flat stretch of land. A developer started building a resort but then abandoned it. Surveying stakes mark the landscaped remains of the course: a golfing graveyard amid fields of grazing cattle.

"That used to be the most beautiful meadow you ever saw," says Ida, her voice wobbly as she wipes her wet eyes.

ON A DAY THREE DECADES past, Bud, his fly rod in hand, sniffed the sage in the air as he studied tree-lined runs and riffles. At Eagle Eye outcrop, a cliff that rises above the upper Colorado River, giant stoneflies with double wings extended flew back and forth like bats. From beneath the water's surface came a boil, a deep stirring. A trout rose, its body taking shape as its head broke the surface to slurp a bug. The fish's dark back curved out of the water and then disappeared.

Bud grabbed at a stonefly the size of a hummingbird that crawled across his neck. A shiver traveled his spine as he peeled this wriggling insect from his skin. He cast his fly and let it drift through an eddy crammed full of fish. A small smacking sound punctured the silence as the fly disappeared in a spot of froth. Bud lifted his rod and felt the pull of something the size of a king salmon. He had never hooked anything so large outside Alaska. Whatever had sipped his fly so politely surged into the current and then sped along the bank. The bulky creature broke off and Bud never saw it. The size of it makes him shake his head to this day. He had attached himself to a mysterious weight that moved through the river.

In the days that followed, he brought plenty of other fish to his net. Rainbow trout gobbled stonefly larvae in the river's depths. And the fish feasted on giant stoneflies that flew clumsily across the water after crawling onto shore and climbing willow bushes to wiggle out of their skin and morph into winged creatures. Female stoneflies skimmed the river to lay eggs, their wings scattering sunlight as they spun like damaged helicopters toward the open jaws of fish. Rainbows gorged themselves until bugs spilled from their stuffed mouths and their bellies stretched round and taut. The fish slashed the river's surface in a frenzy as they fed, sometimes shooting above the water like chrome missiles streaked with red, snatching stoneflies and then crashing back into the river in explosions of spray. A spectacle of gulping mouths and circling fins stirred the upper Colorado in those days.

As the years passed, Bud noticed fewer big rainbows in the river, and the stoneflies too began to disappear. One spring he didn't see a single giant stonefly along this reach near the river's source. In the occasional rainbow Bud now brings to his net, he feels the feeble pulse of the dying Colorado all the way downstream to its delta.

Chapter 4
THE LAST OF THE WILD

Wilderness in the lower forty-eight states is small islands of untrammeled land surrounded by a human sea. In Alaska, wilderness is still the rule, civilization the exception. And the Alaskan weather is unmatched in its ferocity. Many locations claim they have crazy weather—if you don't like it, wait a few minutes and it'll change, the cliché goes. That saying gets shredded in winds that heave in from the Gulf of Alaska; it gets frozen on the tundra when the mercury drops to 50 below.

When you live in Alaska, you learn to rely on yourself. You make do with what you have on hand, and you figure out how to navigate your way through the wilds. Bud says, "You can't count on someone showing up to save your bacon if you get into trouble."

Just back from Vietnam, Bud was living in Alaska and feeling bulletproof. He had jumped from helicopters and survived jungle combat on battlefields riddled with punji sticks and mines. When he heard reports of outstanding steelhead fishing on the Situk River, he organized a trip for him and his wife,

Kaye. Their gear would be dropped off by plane at a remote hut. From there they would paddle a rubber raft downstream to a cabin along the river, where a pilot would pick them up. It seemed straightforward. But trips in the Alaskan bush rarely go as planned.

FEW THINGS SATISFY BUD as much as studying animals. In the backyard of his home in Denver is a koi pond, where fish in shades of gold and orange swim in sinuous lines through the roundness of the pool. They drift across its cobbled floor, their shadows rolling over rocks. They skirt its smooth sides with muscular movements, their delicate fins trailing. "If you really want to get close to nature," Bud tells me as we stand staring into the koi pond, "watch trout working in a pool."

In little streams that spill through Colorado's mountains, Bud has sat on banks unseen by trout and studied them in pools. The scientist in Bud notes how the trout conserve energy by holding in quiet water. When enough food becomes available to justify spending some stored energy, they move into the fast current to hunt insects that float by, filling themselves with calories. Then they fall back into the quiet water to rest. The artist in Bud notes how their colors change as they swim through sunlight and through shadow. He studies how their shapes shift in water that curls around boulders and swirls through chutes. When he draws and paints, his subjects are animals, often trout. He says of watching fish, "There is nothing quite as pure." But before this contemplative watching and sketching came the catching.

Hunting fish fulfills an ancient urge to chase what is slippery and elusive, to capture creatures that inhabit a watery world we can glimpse but never know. It is a world of reflective pools, but also of crushing floods. It flows through us, through every one of our cells, this river that nurtures life. Yet in its waters we can drown.

THE FIRST DAY BUD and Kaye spent on the Situk River after
being dropped off by plane is emblazoned in Bud's memory.
He recalls his young wife sunbathing in her bra on the riverbank while he hunted steelhead in the river's braided flow.
The fish he hooked bulleted into the blue sky and tailwalked
across sandbars, silver acrobats flashing in the sun. The snow
on the surrounding banks had been sculpted by the river into
sparkling cliffs.

Bud kept a few Dolly Varden trout for camp food. That
night, as the smell of grilling fish and the heat from the potbellied stove filled the hut he shared with his wife, who was preg-
nant, he felt they had everything a young couple could want.

When they woke the next morning the blue sky was gone.
The sun had been extinguished by rain that blew in horizontal
bursts. Downpours Bud had witnessed in Asia hadn't packed
as much power as this Alaskan rainstorm.

He considered staying in the hut and not floating downriver
to the next cabin. But Kaye insisted they continue. "We didn't
come this far to sit in the cabin all day," said Kaye. She was
twenty-two years old and had hardly been away from a city, let
alone heading off on adventures in the Alaskan bush, but she
was determined to stick to the plan they'd made. She wanted
to go down the wild river.

Bud wrapped their gear in trash bags and stashed it in their
raft, and they set off, drifting downstream. There were no rapids to navigate, but obstacles more dangerous than whitewater
blocked the way. Snags of trees in a river's weak current might
not seem likely to cause disaster. But when your raft gets stuck
on a snag, and you have to climb out of the slippery craft to
work its rubber sides free from this sharp entrapment, and
the river rises above your hip boots and fills them with chilly
water, and the air temperature is barely above freezing, and
it's raining so hard you struggle to snatch a breath from the
saturated air, you might as well be in a Class 5 rapid paddling
for your life.

Kaye passed from the teeth-chattering and shivering stage of being chilled to the drunken sluggishness of hypothermia. Bud's body heat was prevented from dropping too low by the exertion of freeing the raft from logjams as it ricocheted between roots and trunks. A ferocious wind blasted upstream, slowing their progress, and they made frustratingly little headway. Rain poured down. The stove Bud had brought was too wet to work. Crossing paths with a person who could help them was unlikely. Nor could they turn around and make their way back upriver to the hut where they'd stayed the night before. They were alone in a cold rain that would not stop. Their only option was to keep moving downstream, to keep fighting their way through the wet weather toward the promise of the dry cabin.

All Bud knew of the cabin was that it was an A-frame sited a quarter mile past a cable strung across the river. He squinted his eyes against the slashing rain and paddled forward, into the wind, searching for a gray line across the gray sky, trying to will it into existence.

The cable appeared, a lifeline stretching through the gloom. But Kaye's condition was worsening each drenched minute that passed. Incoherent, she slipped toward unconsciousness. Bud needed to find the cabin and get Kaye dried off and warmed up. He paddled farther downstream, his eyes trained on the riverbank as he searched for a slanted roof among the dim poles of trees.

In the mist-filled forest, he saw the A-frame. He pulled the raft ashore and carried Kaye into the cabin. It was cold inside, but dry. A potbellied stove looked promising. Bud could hardly work his hands. With fumbling fingers, he managed to strike a match alight and touch it to wood soaked with fuel oil—which boomed like a bazooka as it ignited. Bud stuffed more wood into the stove. Then he peeled off his and Kaye's soaked clothes, and together they huddled in two old Army sleeping bags Bud zipped together. They slipped into a deep sleep as their temperatures rose.

In the morning, Bud crawled from the sleeping bag, his body sore and his brain foggy from the ordeal on the river.

He stumbled outside into the bright dawn. The cabin was surrounded by a marsh, its surface shimmering like liquid silver in the sun. Islands of evergreens rose above the flood, and ducks and geese covered the water in a bounty Bud had never seen. The Gulf of Alaska bashed against a nearby beach in a chaos of spray. The river, the flooded forest, the crashing gulf: Water encircled the young couple in a liquid wilderness of currents and waves. Everything around them was saturated, but the storm had blown itself out. Whatever chill remained in their bodies soon vanished in the burning light. Their first child was growing inside Kaye as they waited for the plane to circle overhead and then skim to a landing on ground sheeted with water, the plane's wheels raising silver fans of spray. The world was alive with birds. Winged shapes filled the sky in multitudes, gliding and wheeling before the sun.

THE FIRST TIME BUD WALKED through fields around the Colorado River near the Sheriff Ranch he thought the hay looked as lush as Alaskan meadows. Wind moved in waves of green across the fields, and gusts rose into the trees to rattle the leaves of cottonwoods. He couldn't see any houses or roads. As he climbed over deadfall, a doe and her spotted fawn bounded away. Beaver dams had turned river channels into a patchwork of ponds, and the sky was filled with birds, stirring Bud's memories of the Alaskan wilds.

He walked to the river's edge. With no other person around, no sign of roads or power lines, he watched trout by the dozens hold in the slow curl of eddies, their noses poking from the river as they sipped mayflies. Many of these wild fish had never felt the sting of a hook because so few people came here. It seemed a primeval place, where the river flowed as it had for thousands of years, back to a time when mastodons followed the glaciers creeping northward, and water was released from the silence of ice to sculpt the channels of the Colorado River.

As Bud fed line through the guides of his fly rod, clouds

bunched up in white heaps overhead, dragging shadows across his arms. Sunlight glinted on the water. In the places where the river meandered through meadows it ran swift and clear. The cobbles that lined its bed looked so clean they seemed polished. In a stretch where the river stalled against a curved bank, pools darkened to shades of emerald and jade. And in these timeless pools trout rested and then ripped toward the current, shoving their way into riffles. Water poured across beds of pebbles and sluiced over gravel in currents rich with bugs that were gulped by feeding fish. Some of those bugs were made of feather and fur and were tied to the tippet at the end of line spooled around Bud's reel. And the fish he hooked surged and leaped in the deep green water, and they shimmied through skinny riffles, and he tired them by holding his rod high, and he kneeled in the water and guided them to his net, and he slipped the hooks from their mouths, and he held them in the current until they were strong enough to swim away.

Every spring Bud would watch a spectacle of blue-wing olives in uncountable numbers riding the river's current. The upraised wings of this mayfly species stood tall like the sails of tiny ships. Trout fed on the blue-wing olives in a frenzy of slashing jaws and knifing tails that sliced the water's surface like shark fins as they hunted their mayfly prey.

The blue-wing olive hatch in spring is now a small remnant of what it once was. Some blue-wing olives can still be seen sailboating their way downstream, and a few fish chase after them. But as with many other species on the Colorado River below Windy Gap Dam, blue-wing olives are more abundant in memory than in this degraded waterway. And if steps are not taken to restore this reach of river, the future of many forms of life in the water, and on the banks and beyond, looks bleak.

One day on the upper Colorado, before the river's steep decline, a trout Bud hooked dove deep into a pool and flapped its tail. Fluttering like a silver leaf, it sank toward the bottom, a bright shape dimming in the deepening green. What the Colorado River once was seems a distant dream of the wildness and abundance that filled the world before we corralled its rivers with concrete walls. We still have the great rivers of Alaska that braid across gravel bars and swirl with salmon. Bud would like for us to have an upper Colorado River that swells with spring floods that wash its cobbles clean, restoring the life that once thrived in its waters. He wants his grandkids to turn over rocks in the streambed and see the many creatures that inhabited these waters when he first came upon them. The river is his charge. He understands he won't see it restored in his lifetime. But he is determined to start the healing now so future generations will know some of the bounty this depleted river once held. "I want to leave it better than I found it," he says.

MANY YEARS AGO BUD bought a share of Chimney Rock Ranch next to the Sheriff's spread on the Colorado so he could taste the dust of distant fields, smell the turpentine and citrus scent of sage, and inhale the piney reek of spruce. He would walk for miles through the wilds, listening to wind hiss and grasshoppers rattle as they fled his footfalls, feeling cold water rise up his legs as he stepped through hidden channels.

In the Colorado headwaters, as in Alaska, Bud never knew what he would find. He just knew he needed to follow the river's course, searching every bend, exploring the deepest pools. He didn't go to the river to escape. He went there to spark his senses with all the sharp details of the world that are muted by the haze and clang of city life. Without distraction, he could focus on his favorite bird.

What the American dipper lacks in beauty with its dull gray plumage it more than makes up for in charm. Bud fell in love with these "cute little critters," as he calls them.

White feathers adorn a dipper's upper eyelids, flashing when it blinks. On slender legs the dipper bobs its stocky body up and down. The dipper is drawn to fast-flowing streams, where this aquatic songbird hops between boulders and splashes in pools, its beating wings raising fans of spray. Well equipped for the life of an underwater hunter, the dipper swims beneath the surface to feed on minnows and bugs. An extra eyelid functions like goggles, letting the dipper see as it swims through the water. Special scales pinch closed its nose when it dives.

Bud would sometimes see a dipper emerge from a waterfall, where the bird had built its nest in a nook secreted behind a wall of spray. To watch a dipper dive through the rainbow mist of a waterfall into a crystalline plunge pool is to marvel at the magnificence of the world. To see this stout little bird on stilt legs walk along a streambed while submerged, and then spread its wings underwater and fly beneath the surface, is to be transported back to a state of childlike wonder.

As insect species vanished in the upper Colorado, Bud observed fewer and fewer dippers over the years. Now they have all but disappeared. The once shiny river cobbles lie buried beneath dull silt. The once sparkling riffles choke with weeds. Big fish appear in blurred memories and brightly told stories, but rarely are they seen in the river.

Fly fishers like Bud visit the same stretch of stream week after week, month after month, year after year. Keen observers of aquatic and riparian[6] ecosystems, they witness and document the disappearance of native species like stoneflies, and they call attention to declining numbers of trout. They watch the willows die and the dippers disappear. They see the cottonwoods bleach white as bone and hear their dead limbs rattle in the wind.

When Bud made his case in testimony to state wildlife commissioners that the river was in trouble, he noted his observation that dippers had gone missing. This man who had survived firefights in the jungles of Vietnam and wilderness ordeals in Alaska fought back tears as he spoke of the vanished birds.

Biologists consider the dipper an important indicator species: Its absence signals a serious problem with a stream's

---

6   "Riparian" refers to the banks of a river or other body of water.

ecology. And the Colorado itself is a critical indicator. The consequences of the river's demise ripple outward, affecting the health of the planet.

<center>~~~~~~~~</center>

How can water developers get away with building dams and diversions that degrade a natural wonder like the upper Colorado River, damaging so much life?

In this age of heightened environmental awareness, we are driven to maintain a river's health for the benefit of future generations. We also understand the tremendous economic incentives to protect a resource that sustains a lucrative tourism industry, bolsters the state's image as an outdoor mecca, and performs vital services for free, such as filtering water and forming a buffer against floods.

But Colorado water law grew from mining rushes and an agrarian past, when settling the parched frontier was the first priority. The regulations governing water use throughout the American West are often at odds with what today seems the good sense of preserving a natural resource for the ages. In 2015, killing a river in Colorado is still legal, according to state laws. Perhaps future generations will look back on our slaying of rivers as a barbaric vestige of a vanished past, as brutal as slaughtering elephants for their ivory.

To understand how we arrived at this awkward point when we are still killing the streams that sustain us, we must travel back to the Dust Bowl of the 1930s, when moving the Colorado River across a mountain range made excellent sense.

Part II

Chapter 5
BIG TOM

The little town of Berthoud, its main street tucked between rows of trees on Colorado's northeastern prairie,
boasts one of the best breakfast-anytime diners in the
state. Filled with locals raising coffee mugs to mouths that
exchange the latest news, this cafe would be a mirage rising
from the heat-shimmering plains if not for the Colorado River
headwaters on the far side of the Great Divide.

Each molecule of water redirected from a distant peak to a
faucet on the eastern prairie travels convoluted terrain. And
it's hard to say which is more complex: the infrastructure that
allows people in Grandpa's Cafe in Berthoud to sip water that
began as melted snow on the West Slope of the mountains, or
the history and laws that account for this rearrangement of the
state's rivers and streams.

From a low point of 3,350 feet above sea level where the
Arkansas River trenches through the eastern prairie to the
summit of Mt. Elbert that unfurls banners of snow at 14,440
feet, the geography of Colorado is punctuated by extremes.
The story of the damaged upper Colorado River rests on this

varied physical structure, for water flows where the land allows it to go—or where people force it to go when they try to curb the region's cruel aridity.

From the Pacific Coast, winds heavy with moisture blow eastward, crashing into the Continental Divide and climbing mountain walls. In the cold air of upper elevations, water vapor condenses into droplets and falls as rain or snow on the West Slope. After crossing the Divide, winds wrung of moisture slide down the East Slope, and clouds stingy with rain pass across the plains.

This pattern of prevailing winds and precipitation determines the flow of water in the state. Rivers that run down the West Slope carry an abundance of water; by comparison, rivers on the East Slope contain a paltry flow. Or rather, they did before we reengineered the state's plumbing on a grand scale.

THE LARGEST DIVERSION IN Colorado to move water across the Continental Divide was born in drought and hardship. In the 1930s, as the state suffered though several years of scant rain and the economy languished during the Great Depression, farmers on the northeastern plains were nearly ruined. Rivers emptied, ditches ran dry, and Dust Bowl winds darkened the sky. Irrigators who clung to survival by the thinnest of margins tried to convince Congress to reserve for their use North Platte River water from a federal project being built in Wyoming. But liquid relief was not sent their way.

Desperate for water to supplement the scrawny South Platte and its tapped-out tributaries, civic leaders on the northeastern plains looked across the Continental Divide to the Colorado River. The headwaters form a tantalizing bulge in the mountains. Along the Great Divide, which runs the continent's length, the portion that pushes the farthest east holds the Colorado River's source. Freeing liquid from this wet protrusion for the dry plains below was not a new idea. It was elegant in its simplicity, airtight in its logic.

Farms in the northeast of Colorado, with an abundance of fertile soil and sunshine, boasted some of the state's most productive agricultural lands—but they were in dire need of water. Why not move the West Slope's surplus water to where it was needed and would be put to good use?

West Slope rivers were already being used by fish and birds, but "ecosystem" wasn't yet a word, much less something to save. And economic collapse loomed over the nation. The order of the day was putting water to "beneficial use," defined as using water for industrial, agricultural, domestic, or municipal needs. The rivers of western Colorado would one day support a multibillion-dollar tourism industry spanning several states, but this could not be foreseen by people on the desiccated plains of the 1930s. During the Great Depression, a lucrative business built on catch-and-release trout fishing would have seemed to farmers trying to stave off bankruptcy about as likely as people someday holding devices in their hands that could access all the information in all the world's libraries and could make phone calls, too.

Pioneers who eked out a living on the hardscrabble farms of the prairie rarely had the resources to go picnicking or hiking in the mountains. And whitewater boating in crashing rapids seemed like something a madman or a courageous adventurer might do, not an activity that would fill the state's coffers with tax revenue. A few sage voices spoke of leaving rivers unmolested to run down their ancient courses to the sea. But the words of prophets of preservation such as John Muir disappeared beneath the massive tonnage of concrete poured for dams.

Each spring, the rivers of western Colorado swelled with snowmelt, overflowing their banks and flooding mountain valleys. Only the most fertile imaginations could foresee that redirecting those spring floodwaters beneath the Continental Divide would cause the collapse of the West Slope's ecology. Had any of us been farmers on the plains around Greeley, and had we heard the earth crackle beneath our boots as we walked across drought-stricken fields of dead crops and coughed on soil blown by a brown wind, the mission to move water across the mountains to supplement the South Platte Basin's strained supply would have seemed sensible. Surely we would have signed off on it without the slightest hesitation.

Unless, of course, the engineering challenge of punching a tunnel into the Rocky Mountains to redirect a river through some of the continent's most intimidating terrain struck us as overly audacious. No transmountain diversion of this scale had been built.[7] It would be the largest such project on the planet.

〜〜〜〜〜〜

SETTLERS ON THE PLAINS of eastern Colorado who struggled with stingy local rivers and streams that dried up in late summer and fall had been working to tap the flow of the Colorado River and its tributaries for more than half a century. In 1860 miners managed to move a modest amount of water from the Colorado River headwaters into an upper reach of the South Platte River. And as far back as the 1880s the state had commissioned studies of the feasibility of diverting Colorado River water across the Continental Divide for the eastern plains. Under Colorado's state water laws, these schemes to move water from one river basin to another are legal. Prior appropriation, the foundation of the state's water laws, is so powerful in Colorado it can penetrate the barrier of mountain walls.

The doctrine of prior appropriation means the first diverter to put water to beneficial use has the legal right to continue using that same amount of water. It doesn't matter if the river flows by the diverter's front door, or if it runs down the far side of the mountains. If you can build a pipe long enough to cross a mountain range and put the water to use, you can drink from any river you want, no matter how distant, no matter the

---

7   A "transmountain diversion" (also known as a "transbasin diversion") diverts water from a river's natural drainage basin and sends it through infrastructure such as canals, tunnels, and pumps to another basin on the other side of a mountain range.

consequences to the river or to the human communities along
its banks.

Ever since miners first diverted mountain streams to sluice
boxes that separated gravel from gold, a priority system to use
water formed in Colorado. Those who used water first—usually farmers and ranchers—stand at the head of the line. Then
industry and cities took their place in the lengthening queue.
Organizations like the Sierra Club and Trout Unlimited, which
advocate letting water remain in rivers to preserve ecosystems,
showed up last—and now they occupy the end of a very long
line. Simply put, leaving water in rivers is the lowest priority,
taking water out is the highest.

⁓⁓⁓⁓⁓⁓

BY THE TIME THE DREAM of constructing a massive diversion
system came along during the Dust Bowl, some ambitious
projects to transfer water from the wet side of the Continental
Divide to the dry side had already been built—most notably
the Grand Ditch.

Few words sound less glamorous than "ditch," and adding
"grand" seems an oxymoron: "Grand Ditch" has all the cachet of "beautiful mud." But semantics was not the concern of
farmers who envisioned a far-reaching system to divert snowmelt from the headwaters of the Colorado River (known as
the Grand River at the time). To prevent water in the Never
Summer Mountains from following the lay of the land toward
the west, engineers would redirect it eastward across the Continental Divide at La Poudre Pass—a bold solution to a pressing problem. Delivering water to thirsty farms on the eastern
plains during days of lingering drought demanded a project as
magnificent as the aqueducts of Rome. Built by laborers using
hand tools in the 1890s, this ditch was grand indeed.

Fields nourished by clear gold sent from across the Continental Divide produced an abundance of food, and communities in northeastern Colorado flourished. Water that flows
through irrigation ditches grows more than crops: It sustains

economies centered on agriculture. But when fields are fed by diverted rivers, wild places sometimes starve.

To irrigate farms in Weld County, the Grand Ditch diverts up to 40 percent of the runoff from the Never Summer Mountains. Redirecting the Colorado headwaters into the ditch altered the ecology of the river's upper reaches, reducing the native fish population. But when work on the Grand Ditch began, the sustainability of cutthroat trout in the Never Summer Mountains was, understandably, not a concern.

Anyone who has spoken with a farmer whose property teeters on the brink of foreclosure, whose family's fate is uncertain, can understand why the well-being of critters in a distant watershed isn't what keeps that farmer up at night. And anyone who understands that food doesn't magically appear on supermarket shelves, but requires an enormous amount of work and financial risk, can appreciate the importance of farmers.

The statistic of agriculture being responsible for 86 percent of water use in the state of Colorado is often cited by environmentalists and municipal water managers as if this number were an inherently bad thing. But growing food on Colorado's plains and in its mountain valleys sucks up a lot of water. If Front Range residents want the things they eat to be grown locally, water has to be diverted from the Colorado River. With water in the West, lunch is never free.

In 1915, President Woodrow Wilson signed into law Rocky Mountain National Park, protecting more than 400 square miles of spectacular alpine terrain. But in the eyes of some environmentalists, the park is a beautiful lie because it lacks protection from water projects.

In 2003, the Grand Ditch breached its banks, sending mudslides into the Colorado River headwaters, and damaging sensitive stream and wetland habitat. The National Park Service sued, forcing the company that controls the ditch to pay for ecological restoration. The Park Service has also pushed for reduced diversions in the Grand Ditch to protect the park's natural resources. Surely this would have caused pioneers on plains scabbed by drought to scratch their heads. Leaving water in rivers to benefit fish and wildlife would have struck them as silly at best, dangerous at worst.

The Grand Ditch was followed by other transmountain diversion schemes along the Continental Divide in Colorado. But these projects were modest in scale next to the proposed Colorado-Big Thompson Project—which would send beneath Rocky Mountain National Park a manmade river that would make the Grand Ditch seem a creek in comparison.

Creating a project of this scope would require more than private enterprise. To establish the water infrastructure for irrigated farms to prosper, Westerners relied on the federal government—which they loved to hate, except when it could pay for things they wanted.

The Mormons had managed to establish civilization in some of North America's driest lands. They held an advantage over other settlers because their system of self-government allowed them to oversee the construction of water projects and avoid conflict while managing water distribution. Few other large-scale irrigation efforts could claim long-term success in the American West. (Union Colony, later renamed the city of Greeley, was one notable exception.) Irrigation initiatives had, for the most part, failed. The nation found this unacceptable.

America was determined to fulfill its Manifest Destiny and spread its civilization from shore to shore. The federal government could create the infrastructure for widespread farming in the two-fifths of the country that didn't receive enough rain to grow crops without irrigation. Also, Japan was rising as an imperial power in the Pacific and searching for resources beyond its backyard, prompting politicians to implement a federal program to develop the blank spaces on the map of the American West. Prevailing wisdom held that dams and diversions in arid lands would help build the United States into a world power.

No OTHER FEDERAL AGENCY has had as significant an impact on the American West as the Bureau of Reclamation. Congress created the Reclamation Service in 1902 (later renamed the Bureau of Reclamation) to "reclaim" arid lands for human use. This meant irrigation. To provide a reliable irrigation source, wild cycles of drought and flood, of surplus and shortage, had to be steadied with water storage. This meant dams—lots of really big dams.

Dams built by the Bureau of Reclamation were among the most extensive public works projects in human history. The Bureau tasked its best and brightest with building the infrastructure to make barren lands bountiful. Skyscraping walls in the deserts of the West realized an age-old dream from the dawn of civilization—the control of rivers through human engineering. From the ashes of the nation's collapsed economy in the Great Depression, some of the largest structures on the planet rose. Catastrophic floodwaters were corralled behind towering slabs of concrete. Devastation caused by droughts diminished, and farms and cities flourished. Across the dry wilderness of the American West, the nation solidified its dream of Manifest Destiny in structures of concrete and steel.

Colorado hungered for a piece of the western water development pie portioned out by the federal government. After the sparse waters of the South Platte River and its tributaries had been sucked dry, farmers in northeastern Colorado focused their attention across the Continental Divide on Grand Lake—the state's largest and deepest natural body of water. Delph Carpenter, a Coloradan who spearheaded the Colorado River Compact, an interstate agreement that divides the waters of one of the world's most overused rivers, had dubbed Grand Lake "the last water hole in the West."

In the summer of 1933, the Greeley Chamber of Commerce organized a committee to survey and fund a transmountain diversion to tap the Colorado River headwaters around Grand Lake. Two years later a group called Northern Colorado Water formed to push the ambitious project forward. Composed of civic leaders who referred to the Colorado-Big Thompson

Project as "Big Tom," the group was determined to turn Big Tom from a pipe dream into concrete piping. They saw the project as a means to get people out of breadlines and back to earning a living on farms. To do this, they would need the dollars of the federal government—lots of dollars.

Some critics grumbled that public works dams and diversions were "creeping socialism." But for the most part, federal dollars devoted to water projects in the drought-plagued West were seen as a blessing. Senator Alva B. Adams, a steadfast proponent of Big Tom, relentlessly lobbied the federal government to finance the project—a stance at odds with his fiscally conservative track record of voting against New Deal government spending initiatives.

When water, the ultimate source of wealth and power in the West, could be made to flow through a federal project planned to double potato and sugar beet yields, the libertarian backbone of western politicians became as bendable as rubber. In the unending quest to find new sources of water in the arid West, political ideology has traditionally taken a backseat to pragmatism. The main water priority of politicians in western states historically has *not* been to craft conservative policies that allow a region to live within its limited water means.

To be fair to politicians in the West who espoused libertarian viewpoints while championing federally funded water projects, many of them made a case that the East benefitted from infrastructure paid for by the U.S. government. They argued that because the West contributed greatly to the nation's wealth through extraction of its mineral resources, it should get a generous share of government water projects.

The timing of the proposed Colorado-Big Thompson Project (known as the C-BT Project) was perfect. The federal government under President Franklin Delano Roosevelt was determined to build big public works water projects to create jobs and stimulate the economy. Charles Hansen, editor of the *Greeley Tribune,* became an energetic supporter of lassoing some of those federal funds for northeastern Colorado's farms and communities, earning him the moniker "Godfather of the C-BT."

But there was a catch.

To get the federal government to build the C-BT, Colorado had to present a unified front: Congress was reluctant to sign off on a public works project that would cause conflict within a state. And political leaders on the West Slope did not share the East Slope's enthusiasm for redirecting Colorado River water away from the mountain valleys of western Colorado to the prairie ditches of the Great Plains.

The C-BT was designed to deliver water to irrigated agriculture in northeastern Colorado. But the West Slope wanted to protect irrigated agriculture on its side of the Divide—and it had growth ambitions of its own. The West Slope hoped to someday put more water from the Colorado River and its tributaries to use to support its own urban and industrial growth. West Slope interests staunchly opposed the C-BT, and for several tense years it looked like the holy grail of water development for northeastern Colorado might slip from the grip of East Slope irrigators.

Water wrangling ensued between slopes separated by the Great Divide. Congressman Edward Taylor of the West Slope insisted that if federally funded transmountain diversion projects were to be constructed in Colorado, "compensatory storage" had to be built on the West Slope. This local storage would help make up for the water being diverted—or "stolen," in the heated rhetoric of angry West Slope citizens. When the West Slope puts water to use in its fields and towns, much of that water returns to its rivers and streams. In contrast, water moved through transmountain diversions across the Great Divide is gone for good from the West Slope.

Despite persistent conflict between the East and West Slopes, these Montagues and Capulets of Colorado realized if they didn't put aside their differences and work together to secure federal water projects, their infighting could be exploited by downstream states. The East Slope accepted that to win statewide support for Big Tom, it couldn't just run roughshod over the water needs of the West Slope. And the West Slope, driven by the principle of "my enemy's enemy is my friend," realized rapidly growing California presented an even greater threat than the East Slope. California, by scrambling to make use of Colorado River water and then claiming the right to continue using it, could stymie the state of Colorado's water development.

For Colorado to secure its share of federal funding for water projects, allowing the state to put the Colorado River to maximum use, the East and West Slopes had to quit bickering and get in bed together. They didn't sing "Kumbaya" at their negotiation meetings. But they did avoid clashing in an all-out western water war—which would have made for a great story but a disastrous statewide water policy.

The classic tale of a western water war is Los Angeles's devious dealings in the Owens Valley, immortalized in the film *Chinatown*. Though exaggerated to caricature in the film, the legend of how a bullying city stole water from a rural valley makes for stimulating drama: A cast of villains clandestinely buys up water rights in the Owens Valley and then diverts the Owens River into the bank accounts of sleazy media moguls and unscrupulous real estate developers—leading to the demise of the Owens Valley and the rise of Los Angeles as a metropolitan powerhouse.

Colorado's East Slope found inspiration in Los Angeles's successful construction of its aqueduct that drained the Owens Valley. This engineering achievement boosted the confidence of Coloradans that their C-BT Project could be built. But the way Big Tom's boosters pushed the project to completion would make a tedious film. Meetings between the rival slopes did not erupt in fisticuffs. No trainloads of thugs with tommy guns were dispatched from East Slope cities to rural valleys on the West Slope to wrestle water away from the locals, as in the lore of the Los Angeles–Owens Valley war. No explosions shook the mountain silence as renegades sabotaged the pipes of a distant water empire. Instead, an awful lot of negotiating took place. *Chinatown* this was not.

Colorado's competing slopes forged a peace treaty: a state water policy that provided each side of the Great Divide some of what it wanted. Compromise makes for sound water management but dull stories. Not that drama disappeared completely after the warring slopes reached accord: Additional opposition to the C-BT came from the National Park Service.

Rocky Mountain National Park had been created to preserve the area from exploitation when hunters, trappers, loggers, and miners damaged the land and water in their quest for riches. Park Service officials saw the proposed water diversion tunnel that would pass beneath the park as another form of exploitation; they advocated for an alternate route outside park boundaries. They pointed out that unless specifically authorized by Congress, the Federal Water Power Act prohibited water conduits within any national park or national monument. Groups opposed to the project, such as Audubon and the Izaak Walton League, sent letters of opposition to the Secretary of the Interior.

The project's proponents countered that the act creating Rocky Mountain National Park in 1915 had included an exemption from the ban on water projects in national parks. This was true. So powerful was the irrigation lobby on the East Slope, it had pushed for a provision in the legislation allowing the Bureau to do its reclamation work inside Rocky Mountain National Park. Without this provision, a crown jewel of the National Park system would not have been preserved.

Protecting nature is nice, but water is power. Projects that funnel the most coveted substance in the West toward people with money manage to surmount each obstacle that stands in their way.

The C-BT's supporters made assurances that the surface of Rocky Mountain National Park would not be harmed by the subterranean plumbing. And they claimed the project would be as important to Colorado as discovering gold or building the first railroad. The state's future depended on this diversion of water, they insisted. Charles Hansen, a tireless cheerleader for the C-BT, said National Park officials had "fought like a bunch of wildcats, but I think we have them licked."

Despite opposition from the Park Service and wilderness advocacy groups, the Bureau of Reclamation received permission from Congress to build Big Tom. In the history of the American West, seldom has the principle of wilderness preservation proved as powerful as the need to secure new sources of water.

~~~~~~~~

FOR POLITICAL AND ECONOMIC leaders on the West Slope who fantasized about filling mountain valleys with farms and cities, their dreams of future growth were protected in agreements outlined in Senate Document 80, approved by Congress in 1937. Known as the "Bible of the C-BT," Senate Document 80 has survived for more than three-quarters of a century with a few minor modifications; it guides the project's operation to this day. And if you happen to suffer from insomnia, reading it could provide you with the cure.

The key compromise was compensatory storage. The East Slope agreed that the West Slope would get a storage project of its own to capture water that would otherwise flow downstream beyond the state's borders. Green Mountain Reservoir, located on the Blue River, a tributary of the Colorado, would store water for the West Slope and release it as needed to make up for Colorado River flows sent across the Continental Divide. This compensatory storage worked well for the water needs of West Slope communities; it didn't work as well for the river itself.

The agreement wasn't intended to compensate the West Slope's environment—and could even be taken to have the opposite effect. For every scoop of water the East Slope removed from the Colorado River, a second scoop was stored in a reservoir on the West Slope, instead of flowing freely down the Colorado's course toward the sea. The Colorado was becoming dead plumbing rather than remaining a living river. Some environmental safeguards were, however, written into the legislation that authorized the C-BT.

Colorado's past water leaders often suffer the fate of being maligned by modern environmentalists as megalomaniacs determined to destroy nature. But some of their actions can seem forward-thinking to contemporary sensibilities. To appease proponents of protecting Rocky Mountain National Park and its surrounds, Senate Document 80 required the project be operated "[t]o preserve the fishing and recreational facilities and the scenic attractions of Grand Lake, the Colorado River, and the Rocky Mountain National Park."

Whether water providers and government agencies have followed this mandate to manage the C-BT without harming the river is another matter. Nevertheless, this is a clear statement of the importance of preserving the natural environment, immortalized in a document written in 1937 by water leaders who were, by and large, honorable people doing the best they could with the information they had, and operating within the value system into which they'd been born. While the nation was charging forward as fast as it could to develop the rivers of the American West, the document's authors deemed protecting a stretch of the Colorado important enough to feature amid provisions that would siphon water to distant farms on the droughty plains. The environmental protections in Senate Document 80 may strike us today as toothless. But that they were even included is notable given the spirit of the times, when the nation was teetering on the edge of economic catastrophe, and waterless farmland was blowing away beneath the feet of people whose stomachs groaned with hunger.

Now our supermarkets overflow with food, and the upper Colorado River is a starved skeleton of what it was in 1937. To people like Bud Isaacs who are determined to return the Colorado headwaters to health, the directive to preserve the river in Senate Document 80 confirms the rightness of their cause.

Chapter 6

WATER IS POWER

The year 1937 was a watershed in water management. Along with creating Senate Document 80, Colorado enacted state legislation establishing a framework for "conservancy districts" and "conservation districts."

"Water conservation" meant something much different in 1937 than it does now. Conservationists opposed leaving water in rivers to serve no practical purpose; they made sure every drop was put to human use. The U.S. Postal Service once issued a stamp with an image of a dam and the word "conservation." The C-BT was conceived as an ideal water conservation project. Today, most people would consider a system that dries up a river's natural flow the opposite of conservation.

Colorado's 1937 legislation laid out a plan to finance and oversee the C-BT. It also added words to the ever-expanding lexicon of Colorado water terminology: an overabundance of concepts and language that ensures water attorneys will never be bored—and ensures citizens with limited attention spans will feel overwhelmed by the absurdly complicated legal system that manages such a simple substance.

The Northern Colorado Water Conservancy District was established in 1937 as the first water conservancy district in the state. *Who cares?* you might ask. Well, this entity was tasked with working with the federal government to contract for, and then build and operate, the C-BT Project—and the C-BT is at the heart of why the Colorado River headwaters are dying.

Fortunately for the attention-challenged student of water history, the Northern Colorado Water Conservancy District is usually referred to as Northern Water—or just Northern. It's hard to imagine an uglier acronym than NCWCD. The state of Colorado may someday need to institute a Department of Hideous Acronyms (DHA).

Northern Water got to work securing approval of residents for a property tax to repay the federal government a portion of the C-BT's costs. Interest on water projects was forgiven according to the federal Reclamation Act, passed in 1902. Interest on forty-year repayment plans can exceed the principal many times over—though this tends to be overlooked by politicians of the pull-yourself-up-by-the-bootstraps ideological bent who scramble to bring federal water projects to the regions they represent.

With the country mired in the Great Depression, public works projects such as Hoover Dam and the C-BT made sense for America. The nation needed a massive economic stimulus. It also needed a shot of self-esteem. The C-BT wasn't as uplifting as Hoover Dam to the spirits of a downtrodden nation. It was, however, another showpiece project to demonstrate to the world America's engineering excellence and admirable work ethic.

Questions about the fairness of redistributing tax dollars for water projects in the American West weren't as vexing when the C-BT was built as they are now, when we learn that subsidized water is used to grow subsidized sugar beets in a system that looks suspiciously like something the Soviets might have created. The twin federal subsidies of water projects and commodity support payments continue to make sugar beet farming in northeastern Colorado profitable for a fortunate few. Why a taxpayer in Tennessee or New Jersey should cough up part of his income for farmers in northeastern Colorado to grow sugar beets seems more troubling today than it did in the 1930s.

But acts of Congress have been passed. Laws are in place. And leaving water in rivers to support the health of ecosystems, instead of diverting it to irrigate a crop that earns growers entitlement payments through an outdated commodity support program, is no simple feat. Creating a subsidy is as easy as letting water run downhill. Removing a subsidy is as challenging as pushing water back uphill.

Convincing Congress to pay for the C-BT was a cinch. The Bureau of Reclamation, the agency tasked with overseeing the construction and operation of the project, craved building big, complicated projects; its engineers were hungry to take on the technical challenge of Big Tom. But first the Bureau had to demonstrate the financials were sound—arguably a feat more daunting than constructing the project.

A commission chaired by Herbert Hoover in 1955 determined that the ninety projects the Bureau of Reclamation was building at the time would cost twice their original estimates. The cost projections that Bureau accountants came up with convinced Congress to authorize dams and diversions. And once contractors began pouring concrete, a water project was followed through to completion, even when the true expense could no longer be ignored. If you were to think a water project authorized by the federal government couldn't run four times over its original budget, you would be wrong. Very wrong.

Before construction on the C-BT began, the Bureau estimated the cost at $44 million. Congress signed off on the project based on this figure. Why question the agency that had built so many wonders of the modern world? The Bureau had competently constructed Hoover Dam and other marvels. Engineers who could raise skyscraping structures to hold back floodwaters and harness the power of rivers surely were good with numbers, right?

Bureau employees were indeed geniuses at engineering—their technical expertise was beyond dispute. Accounting, well, that was a different story. The project's final price tag: $162 million. Imagine what would happen to a private company that strayed so far when projecting and controlling costs. In the Bureau's defense, World War II slowed the pace of construction, and inflation sent costs spiraling.

Even more striking than the quadrupled cost of the C-BT is that Northern, before construction began, negotiated a contract with Congress to repay a fixed $25 million portion of the project's cost *without interest* over forty years. In effect, the cost of the C-BT was covered by a dentist in Dayton, a tire salesman in Detroit, a plumber in Topeka, a banker in Boston, and millions of other people across America for four decades. A C-BT brochure published by Northern says the repayment scheme is "regarded by many as a great bargain for Northern Colorado residents." If there were an Olympics of understatement, that sentence would be a strong contender for gold.

As mentioned, not charging the beneficiaries of a reclamation project interest on their repayment obligation was written into the 1902 Reclamation Act. Intended to turn barren lands into bountiful family farms, the legislation banned agricultural operations larger than 160 acres from using water from reclamation projects. Many farms that planned to benefit from the C-BT were much larger than the 160-acre limit. This posed a problem. But similar to how cost constraints seldom stopped dam building in the arid West, laws rarely stood in the way. Troublesome laws could be rewritten.

The C-BT was designed to provide supplemental water to farms that already existed—a first for the Bureau. Every other reclamation project had been designed to open new lands to farming. Senator Adams of Colorado, a champion of preventing the federal government from meddling in the affairs of states, introduced a bill to exempt from the land acreage limitation farms serviced by the C-BT. The Senate passed his bill, clearing the way for a federally financed tunnel that would one day bear his name. This subsidized structure diverts water toward the fields of farmers dogged by erratic rain. And those farmers pay pennies on the dollar for the water.

What do we make of this generous gift to a fortunate few, this massive redistribution of water and wealth? To call it welfare doesn't seem fair: The farmers who benefited from the C-BT have a difficult and important job. But so do miners and nurses and diesel mechanics. Why would the federal government take on the role of moving water between regions, depriving a place of its most precious resource to benefit farmers somewhere else? How did the most conservative states in the nation embrace a program as close to socialism as this country has ever come? Why were politicians in the West expected by their constituents both to criticize the federal government for meddling in their affairs and to convince the government to finance water projects they couldn't build themselves?

Reclamation in the West was driven as much by emotion as by logic. New Deal policies that promoted water projects like the C-BT were intended to create jobs and stimulate the economy, but a major impetus for the reclamation movement goes back much further—all the way to the Old Testament. The Bible is replete with powerful imagery of bringing water to desert lands to make them "bloom as a rose." Proponents of government schemes to irrigate the West made frequent use of this symbolism. William E. Smythe, one of America's most prominent irrigation evangelists at the turn of the twentieth century, published a book titled *The Conquest of Arid America*. He viewed irrigation as "nothing less than the progenitor of civilization in an otherwise inhospitable land—the key to making the desert bloom."

Reclamation was as much a utopian social program as a practical effort to promote farming. Greening the American West became a moral cause. Consider the words of Franklin Lane, secretary of the interior under President Woodrow Wilson: "The mountains are our enemies. We must pierce them and make them serve. The sinful rivers we must curb."

Anxiety gripped America when the industrial age dawned in the early twentieth century. As crime and poverty ravaged rapidly urbanizing areas, the nation headed toward what seemed a dystopian future, and people pined for an arcadian past.

Thomas Jefferson had judged cities as "pestilential to the morals, the health and liberties of man." The reclamation movement was inspired by the Jeffersonian ideal of farmers on small plots developing sound moral values as they cultivated crops. Politicians prescribed farming as a cure for the social malaise caused by cities; they insisted morality could be restored if citizens went back to the land. William Kahrl wrote in his book *Water and Power* that proponents of irrigation saw it "not only as a means of economic development but also as the driving wheel for social and spiritual progress."

This drive to achieve social and spiritual progress led to damming damn near every river in the American West, most of them many times over. The reclamation movement continued to gather force and reshape the arid West until a new utopian movement, environmentalism, supplanted the idealization of irrigation with a radical new vision of river management.

But before enviros shut down the Bureau of Reclamation's bulldozers and silenced its cement mixers, federally financed dams in the West bolstered America's economy and confidence. Beginning with FDR's New Deal, dams turned parched landscapes into some of the planet's most productive farmland, providing a level of food security unprecedented in human history. Hoover Dam helped lift the nation's downtrodden spirits, and massive dams and diversion systems allowed dusty settlements like Los Angeles, Phoenix, and Denver to drink from distant rivers and develop into thriving cities. The hydroelectricity produced by Columbia River dams was essential to manufacturing aluminum for the airplanes that defeated Germany and Japan in World War II. These monumental structures also produced the energy used to create the atomic bombs dropped on Japan. The war that shaped civilization so profoundly in the twentieth century could have taken a different course if not for federal efforts to build colossal dams on the Columbia.

It's easy to bash America's dams from the comfort of our safe and well-lighted homes, which exist, in large part, because of the dams we now love to hate.

WHEN THE C-BT WAS authorized by Congress in 1938, the massive amount of federal spending necessary to force water out of the Colorado River and into irrigation ditches on the other side of the Continental Divide was seen as more than a gift to a few sugar beet farmers. The U.S. government paid for the project under the premise that the C-BT was of national importance. The project would help support a farming economy on the droughty High Plains so civilization could thrive from shore to shore, people in breadlines could get back to work, and America could rise from the doldrums of the Great Depression.

When World War II loomed on the horizon, the project's proponents made the case that building the C-BT was a matter of national security because it would increase the country's food supply. But when water was first diverted through the partially completed project in 1947, the war was over, and the water grew the profits of farmers on the northeastern plains of Colorado.

Like Shakespeare's wise fool, a person unschooled in the ways of water management in the West and lacking a law degree could ask some straightforward questions: Because the C-BT was funded by American taxpayers, does the agency that operates the project owe an obligation to the nation at least equal to, if not superior to, its responsibility to provide water to citizens of one corner of Colorado? If a majority of the American people prefers that some of the water diverted by the C-BT to the Front Range instead remains in the Colorado River to restore the watershed's health, could the project be operated in this way? Would the nation benefit more from subsidized sugar beets and suburban sprawl on the Front Range or from healthy Colorado River headwaters?

When American taxpayers footed the bill for space exploration programs, they were repaid in dividends both obvious and indirect. Few fiscal conservatives would make the case that President Kennedy's project to land a man on the moon was a poor use of a few cents on each taxpayer dollar. The mission inspired a generation of Americans to explore the frontiers of science, and practical spinoffs from NASA's discoveries range from Doppler radar to the Space Shuttle's docking algorithms applied to LASIK eye surgery.

When taxpayers picked up the tab for the C-BT, what did they receive in return? Certainly not Doppler radar and LASIK surgery. First they got potatoes and sugar beets. Then they got shopping malls.

Taxpayers were made to open their wallets for a project with the main purpose of providing supplemental irrigation water to farms on Colorado's northeastern plains—this is clearly stated in Senate Document 80. The project did boost the nation's food supply. But when the Great Depression ended and the post-war population boom began, much of the water delivered by the C-BT was transferred from agricultural use to municipal use. The American taxpayer had been billed for a project ostensibly for agriculture—but which soon supported suburban sprawl. Critics of the CBT see this as a clear case of mission creep.

Front Range developers owe a great debt to the C-BT and, by extension, to the federal government. Without a steady flow of water from the heavily subsidized C-BT, the subdivisions that have replaced farmland along Colorado's northern Front Range would be as ephemeral as the rains that vanish in the dry air before they touch the plains.

Of course, government subsidies that support highway construction and home mortgages have also helped fuel rapid growth along the Front Range. A region's available water supply doesn't control the rate of growth, but the way water is distributed and managed can determine *how* a region grows—by either stopping or stimulating sprawl. One can look at the ill-conceived communities along the northern Front Range that drink from the C-BT and see that the project has been used as a tool to support unsustainable development—a tool paid for by U.S. taxpayers and administered by an agency unaccountable to those taxpayers. Northern Water has made life easy for sugar beet farmers and real estate developers, but difficult for dippers and trout of the Colorado River headwaters—and for people trying to protect this resource from ecological ruin.

But the story is getting ahead of itself. Before developers peeled up prairie sod on the eastern plains to pour foundations, and before gridlocked traffic clogged Front Range roads, the C-BT had to be constructed—no small task. Big Tom isn't a jaw-dropping structure like the giant dams of the West. But what it lacks in sheer size it more than makes up for in staggering complexity.

THE C-BT IS A TESTAMENT in countless tons of concrete and steel to America's engineering excellence. It was our moon mission in a time before rockets. It was a national triumph in an age when controlling the American Nile that raged in bouts of flood and drought formed the West's final frontier.

Water megaprojects inspired a generation of engineers with dreams of taming rivers and harnessing their energy—a thorough accounting of the C-BT must be cognizant of this. The power generated from the project's hydroelectric plants must also be considered: A portion of the power revenues helped repay the cost of constructing and operating the dams, tunnels, and pumps. Clever dam designers found ways to make the Colorado pay for its confinement.

Water diverted from the Colorado River through the C-BT is first stored in Lake Granby[8] on the West Slope. From Lake Granby, the water is pumped into Shadow Mountain Reservoir, where it flows by gravity into Grand Lake and then

---

8   Lake Granby is a reservoir formed by a dam, not a natural lake. Grand Lake is a natural body of water.

passes through a tunnel beneath the Continental Divide in Rocky Mountain National Park. The tunnel is 13.1 miles long, the exact length of a half marathon. The redirected West Slope water emerges from the dark tube of the Alva B. Adams Tunnel into the light of the East Slope at Estes Park, where it merges with the Big Thompson River—which becomes much bigger with this added flow. The diverted water then cascades down mountain slopes, passing through six power plants to generate hydroelectricity. In reservoirs on the plains below, the hard-working water rests before being routed through pipes and canals to the fields and faucets of the Front Range.

If this system is confusing to conceptualize, imagine trying to build it. Constructing the pyramids was straightforward by comparison: square base, pointy top, no problem.

After completion in 1957, the C-BT Project moved some 230,000 acre-feet of water annually from the Colorado River Basin to the South Platte River Basin. It's worth pausing a moment to explain that an acre-foot is a standard measurement used by water wonks that doesn't mean anything to the average person. One acre-foot of water is generally considered enough to meet the needs of two families for one year. But that's still pretty abstract. To wrap your mind around 230,000 acre-feet, think of 230,000 football fields each filled a foot deep in water. Or, better yet, think of a tower of water with a base the size of a football field rising more than forty miles into the sky—almost eight times the height of Mount Everest. This tower of water would soar through the troposphere, where all the weather and clouds are contained, rise through the stratosphere, and then climb into the mesosphere—to more than four times an airplane's maximum altitude. That's how much water the C-BT removes each year from the headwaters of the most endangered river in America. The amount increases considerably when the Grand Ditch and Denver Water's diversions are added. And efforts are underway to remove even more.

Building the C-BT infrastructure that helped make this massive diversion of water possible was a monumental feat—perhaps more impressive than raising Hoover Dam.

Tunneling crews worked from opposite sides of the mountains to puncture the rocky backbone of the continent. When the final dynamite blasts joined the two tunnel halves in an excavation half a marathon long, the error in alignment was a mere fraction of an inch. To call Big Tom a manmade wonder of the world is not hyperbole.

Though a laudable engineering achievement, the C-BT was also a "massive violation of geography," noted the historian David Lavender. Water destined by nature's design to flow to the Colorado River Delta and into the Pacific Ocean was rerouted toward the Mississippi River in the Atlantic watershed. The clean kilowatts of renewable energy generated by the C-BT light up East Slope shopping malls with water that would have passed through the gills of fish in the Colorado River. Instead, the water passes through a tunnel named after Senator Alva B. Adams, a fiscal conservative who convinced the federal government to redistribute water and wealth. A testament to political pliability in the quest to secure water lies buried nearly 4,000 feet under the continent's crest, in the silent heart of the mountains. Beneath the pristine peaks of Rocky Mountain National Park, below its clean rivers and sparkling lakes, a tunnel lined with concrete tells the story of our civilization's efforts to reengineer the flow of water on a stunning scale. The Adams Tunnel is a monument to human ambition—and to the dark consequences of our relentless quest to rearrange nature.

***

As PLANNED, A STEADY supply of C-BT water supplemented unreliable flows in the South Platte Basin, boosting northeastern Colorado's agricultural productivity. In the drought-ridden growing season of 1954, water delivered by the project nurtured more than half the $41 million worth of crops harvested that year. In addition, the hydropower generated by the project helped the state meet its energy needs.

The project's reputation spread, drawing dignitaries from nations across the world to Colorado to learn how their own engineers and bureaucrats could mimic the C-BT's construction and operation. Even Marc Reisner, the Bureau of Reclamation's most persistent faultfinder, praised the C-BT in *Cadillac Desert,* his groundbreaking screed against water projects in the West:

> One of the Bureau's most successful projects, Colorado-Big Thompson, was already delivering Colorado River water across the Continental Divide through a tunnel to the East Slope; the power produced by the steep drop down the Front Range was enough to justify the expense of the tunnel, and the additional water diverted from the upper Colorado to tributaries of the Platte River was welcomed by everyone from canoeists to whooping cranes to irrigators in Colorado and Nebraska.

Though Reisner was a perceptive critic of schemes to move water between river basins, in the case of the C-BT he seems to have missed a painfully obvious point. While water delivered by the project allowed the economy of northern Colorado's Front Range to grow and helped whooping cranes prosper in the Platte, the upper Colorado River, starved of its native flows, was under siege.

WHEN WORK BEGAN ON the C-BT in 1938, the first component built was Green Mountain Reservoir. As mentioned, the reservoir was designed to store water within the Colorado River Basin and release it as needed, compensating the West Slope for flow diverted to the East Slope. In the overall accounting of statewide water distribution, Green Mountain Reservoir helps balance the books. But the reservoir is sited on the Blue River, a tributary that empties into the mainstem Colorado nearly forty river miles downstream from where the C-BT diverts water. This creates the infamous "hole in the river."

If you are a water user below Green Mountain Reservoir, the protocol that guides its releases works well for you. But if you own land along the Colorado River above Green Mountain Reservoir, the water stored in the reservoir does you no good. The C-BT diverts flow upstream, draining the river that runs through your property. Green Mountain Reservoir refills the river downstream of you. And you are stuck in the middle—in a hole in the river in danger of running dry.

Such is the fate of Chimney Rock Ranch that Bud co-owns. The ranch is sited below the C-BT diversion but above Green Mountain Reservoir. It lies in the forgotten stretch of river that bears the full brunt of diversions to the Front Range.

But the main offender in turning Bud's section of the Colorado from a healthy river teeming with trout to a sluggish flow stripped of life is a dam called Windy Gap.

Chapter 7

THOSE DAMNED DAMS

The Colorado-Big Thompson Project helped slake the thirst of farms in northeastern Colorado—but when the project was completed in 1957, the Front Range urban corridor was in a post-war boom. Cities slurped up their portion of C-BT water and grew desperate for more. The overused South Platte River Basin in their backyard offered no relief. Some agricultural water was transferred to cities, but not nearly enough to support explosive population growth on the northern Front Range.

Converting more irrigation water to municipal use seemed a bad option. The region's economy relied on a strong agricultural sector, and its quality of life depended on farmland—wedges of open space kept communities from colliding in continuous sprawl.

So in a case of déjà vu all over again, the cities of the plains set their sights across the Great Divide. Northeastern Colorado once more began coveting the Colorado River. And in 1967, the mayor of Longmont, acting on behalf of a coalition that included six cities—Longmont, Loveland, Fort Collins,

Greeley, Boulder, and Estes Park—filed for water rights on the Colorado and proposed a diversion project at a place called Windy Gap.

State laws allowed for Colorado River water that wasn't already being put to use to be diverted outside the basin to distant cities. But the public's concerns had shifted since settlement of the frontier, when prior appropriation had served as the foundation of Colorado's water laws. In the 1960s, interest rose in leaving water in rivers for the benefit of fish and birds— and for people with binoculars and boats, with backpacks and fishing rods.

When the Windy Gap Project was proposed, longtime West Slope residents hostile to transmountain diversions, which they tended to view as the East Slope thieving their water, were joined in their resistance by a new breed of Westerner—environmentalists. And the project's momentum skidded to a stop.

THE ENVIRONMENTAL MOVEMENT grew into a force powerful enough to alter water policy in the West by reacting to a scheme to dam Echo Park. Located in northwestern Colorado at the confluence of the Green and Yampa Rivers in Dinosaur National Monument, Echo Park was the site of a proposed dam authorized by the 1956 Colorado River Storage Project Act. Because the dam would have inundated canyons of the national monument, the plan instigated a clash between the Bureau of Reclamation and the Sierra Club so loud it reverberates to this day.[9]

Sierra Club founder John Muir had failed to spare Hetch Hetchy Valley in Yosemite National Park from drowning. Before the valley disappeared behind a dam completed in 1923,

---

9   The storied conflict between Floyd Dominy, the brash Bureau of Reclamation leader who championed Echo Park Dam, and David Brower, the shy mountaineer who spearheaded the Sierra Club's opposition, was immortalized by John McPhee in his book *Encounters with the Archdruid*.

Muir wrote of Hetch Hetchy, "no holier temple has ever been consecrated by the heart of man." A few decades later, the Sierra Club's Echo Park campaign woke the public up to what was at stake in this stone temple enclosed by soaring walls. Was the nation willing to trade its remaining places of wild beauty for more water projects?

The public railed against Echo Park Dam, and against two proposed dams on the Colorado River in the Grand Canyon. The dams were cancelled, and snowballing hostility toward water projects inspired the 1964 Wilderness Act, protecting many of the undammed landscapes and waterscapes secreted away in the last blank places on maps.

In the 1960s and '70s, Cleveland's Cuyahoga River catching fire, along with Rachel Carson's book *Silent Spring* and other incendiary investigations of industrial civilization, fanned the sparks of environmentalism into a blaze. Citizens demanded the right to participate in water management so they could protect river ecosystems, and a wave of federal laws to preserve the natural environment swept the nation. Along with the Wilderness Act came the Wild and Scenic Rivers Act, the Clean Water Act, the Endangered Species Act, and several other far-reaching pieces of legislation that changed the federal government's management of rivers.

These laws slowed the pace of water development as scientists studied in detail the impacts of building new dams. Gone were the days when a project was proposed and then swiftly built by the Bureau of Reclamation or the Army Corps of Engineers. The federal government's role started to flip from financing and constructing water projects to safeguarding rivers from environmental harm caused by dams and diversions.

In Colorado, the practice of starving West Slope rivers to feed population growth on the East Slope came under increased scrutiny. Paradoxically, people flocking from across the nation to the sprawling suburbs of the Front Range questioned the wisdom of sacrificing healthy watersheds to fuel development. Citizens realized that the exceptional quality of life in Colorado could be diminished if the state's rivers were

drained. Many wondered what good was a home with a garage and a green lawn if on weekends you went to the mountains and found dead fish in a shriveled river. For increasing
numbers of Coloradans, the American dream spread beyond
the fences of their suburban yards to encompass places of wild
beauty in need of protection.

In 1973, the state of Colorado created an "instream flow
program." This was an exception to the water right requirement of diverting flow out of rivers to satisfy human needs.
With the instream flow law, leaving water in rivers to support
fish and wildlife and healthy ecosystems became a recognized
water right. Some experts argue this shift shows the flexibility
of the state's water law system, which can bend to accommodate the public's changing concerns. Critics of the program
counter that it doesn't ensure healthy rivers.

Stream ecologists explain that to build a sturdy house, first
you must secure a solid foundation—a liquid foundation, in
the case of a river. Without maintaining instream flows robust
enough to support aquatic and riparian ecosystems, other environmental safeguards are meaningless. A river needs to be
protected from pollution, but more than anything else, a river
needs water. As a house with a flimsy foundation will fall, so
an ecosystem with insufficient flow will collapse.

Instream flow rights are almost always junior rights. During
a drought, senior diverters keep diverting, junior instream
flow rights aren't fulfilled—and streams suffer. The Crystal
River is a poster child for the sad inadequacy of the state's instream flow program. The river is "protected" by an instream
flow right, but in drought years the Crystal can turn as dry as
a lecture on water policy, while senior diverters flood emerald
fields of hay.

Renegades outside the water establishment point out that
the state's system of water governance was codified when slavery was legal. And they emphasize that even the most modest
efforts to protect rivers encounter determined opposition from
powerful interests. Suggesting to the Colorado water establishment that the state's water laws are in need of an overhaul

is like telling a fundamentalist Christian the Bible should be rewritten.

And so the state's water wars rage on. Guns are safely holstered, but hostile words are fired back and forth between two camps: those who believe Colorado's water laws are sacred and can bend but must not be broken, and those who say scrap the whole wicked mess and start over with a legal system that makes protecting the health of rivers paramount—a system more in step with the twenty-first-century values of river preservation than the nineteenth-century values of resource development.

PUBLIC PUSHBACK AGAINST DAMS and diversions costly to both the national treasury and the natural environment came to a head with the Animas-La Plata Project (add "A-LP" to the ever-growing list of acronyms). After being authorized in the 1968 Colorado River Basin Project Act, the A-LP sparked public debate as Americans wondered why they should pay for this elephantine scheme to deliver water to hay farmers in a remote corner of Colorado. Fiscal conservatives joined environmentalists in ridiculing the A-LP as "a billion-dollar boondoggle." After repeated delays caused by cost overruns, construction almost began in the early 1980s.

But the A-LP made President Jimmy Carter's "hit list," which targeted for cancellation water projects he thought would cause environmental damage and would have questionable economic benefits. Carter's anti-dam crusade, though disastrous for him politically, marked a critical crossroads in federal dam building. Subsequent administrations adopted Carter's stance, navigating the nation toward a more restrained water development policy.

Beginning with FDR's New Deal water projects, the food security and urban development that had resulted from aggressively damming the Colorado and other rivers of the West allowed the United States to build itself into a global powerhouse. But President Carter joined an increasing number of American citizens in pointing out that these dams also had tremendous costs.

Federally financed water projects led to profligate spending that gobbled tax dollars. And dam megaprojects put giant agribusinesses on the public dole instead of fostering small family farms, as the 1902 Reclamation Act had intended. The Southern Pacific Railroad—the largest private landowner in California—became the biggest beneficiary of the Central Valley Project, the most extensive water project in history. After the Bureau of Reclamation rearranged the rivers of the Central Valley, corporate giants such as Exxon and Getty Oil ran megafarms watered by taxpayer dollars. The original intent of the Reclamation Act to support small family farms warped into water welfare for the rich.

And there were environmental costs. After the nation's dam-building binge reinvented the watersheds, farms, and cities of the American West, only a handful of rivers, such as the Yellowstone and the Yampa, still flowed free. Several endangered species hovered on the brink of extinction. And researchers shined the light of science on the dark downsides of dams, such as silt buildup in reservoirs and evaporation of stored water. Damage to the natural world turned public opinion against water projects, climaxing in the controversies over the proposed Echo Park and Grand Canyon dams.

Because the national mood had swung from supporting big water projects to opposing them, the backlash that slapped President Carter after he issued his anti-dam hit list caught him by surprise. He had failed to heed the timeless formula: Water is power. By attacking water welfare projects, this sweater-wearing environmental warrior picked a fight with some of the nation's most powerful interests.

Carter stated, "In the arid West and across the entire nation, we must begin to recognize that water is not free—it is a precious resource. The cornerstone of future water policy should be conservation and wise management."

But the water establishment in the West took conservation and wise management to mean building more and bigger dams with taxpayer dollars. Politicians traded favors by funding each other's dams, generating windfalls for the constituencies that elected them. Former Bureau of Reclamation Commissioner Dan Beard claimed that pork-barrel water projects played such a prominent role in the legislative process, the Civil Rights Act would not have passed without them.

Never having served in Congress, Carter underestimated how deeply entrenched dams had become in the nation's political landscape—and he paid a steep price. His campaign to curtail the federal dam-building bureaucracy in the West was arguably as much a factor in his one-term presidency as the Iran hostage crisis.

When it came to water policy, President Ronald Reagan was Carter in a cowboy hat. Reagan, guided by his principle of fiscal conservatism and his campaign promise to slash the federal budget, enacted a cost-sharing plan similar to one Carter had proposed, requiring states to help foot the bill for water projects. Also like Carter, Reagan threatened to veto financially unsound dams. The Bureau of Reclamation, the agency that had completed many dozens of the most impressive construction projects on the planet, powered down its bulldozers and concrete mixers. And the era of federally financed dams more or less came to a close.

Yet the Animas-La Plata Project, described by *U.S. News & World Report* as the "last surviving dinosaur from the age of behemoth water schemes," just wouldn't die. In 2012, more than four decades after Congress authorized the gargantuan project, the Bureau of Reclamation completed a scaled-down version of the A-LP.

Nor did the environmental community stop the Fryingpan-Arkansas Project, designed to tap Colorado River tributaries for irrigation in the Arkansas Valley and for municipal use in Pueblo, Colorado Springs, and Aurora. Known as Fry-Ark, this sprawling system continued to reengineer the West's hydrology. At a 1962 ceremony in Pueblo, President John F. Kennedy dedicated Fry-Ark, declaring the project "belongs to all the people of the country." Many of those people, however, were advocating letting rivers run in their natural courses.

Completed in 1981, Fry-Ark was the southern Front Range's answer to the C-BT. But coming several decades after construction of Big Tom, and built not in an era of Dust Bowl desperation but in a time of rising environmental concern, Fry-Ark looked to many Coloradans less like an admirable engineering achievement than a convoluted scheme with grim consequences for tapped-out rivers and streams.

Like the C-BT, Fry-Ark's primary purpose was to divert flow from the Colorado River watershed to supplement irrigation on the eastern plains. But much of the water delivered by the project for farming was soon converted to municipal use to support sprawl. Fry-Ark helped change the state's circulatory system, allowing Front Range cities to metastasize as they slurped up more water.

People on the Front Range wanted unsullied nature. They also wanted unlimited growth. How these conflicting desires play out in coming years will, to a large degree, determine the fate of one of the planet's most imperiled rivers.

THOUGH UNABLE TO QUASH Fry-Ark and the A-LP, the environmental community achieved success on other fronts—most notably in the epic fight over Two Forks.

Two Forks Dam was a proposed $1 billion structure on the South Platte River that would tower more than 600 feet—a project that harkened back to the days of Hoover Dam, but with a major difference. The city of Denver and its suburbs planned to pay for and build the megadam without help from the federal government. But the financing plan didn't stop bitter legions from battling to defeat the project.

Two Forks would send water to the Denver metro area so toilets could flush and lawns could grow. It would also, critics contended, deprive wetlands downstream in Nebraska of their lifeblood, damaging habitat that nurtured endangered whooping cranes, along with sandhill cranes and other migratory waterfowl. And the dam would drown Cheesman Canyon, wiping out a world-class trout fishery and eradicating thirty miles of scenic South Platte River. The director of Colorado Trout Unlimited, a nonprofit conservation organization that spearheaded opposition to Two Forks, referred to the reach of river the reservoir would defile as "our holy water."

Citizens appalled by sacrificing a wild sanctuary for yet another reservoir protested en masse. The groups that fought Two Forks made the case that the project was too expensive and too environmentally destructive. With a rigorous approach that relied as much on crunching numbers as on provoking emotion, they showed that conservation, efficient use, and smaller supply projects could fill Denver's water shortfall.

In 1990, as the roar over Two Forks echoed in the halls of Washington, the Environmental Protection Agency silenced the conflict by vetoing the dam. Dr. Gene Reetz, an EPA employee who played a pivotal role in the veto process, explained to me that by law the Army Corps of Engineers—the agency in charge of issuing the permit for the project—had to choose the "least environmentally damaging practicable alternative" for solving the Denver metro area's water shortage. Because the dam's adversaries had demonstrated ample opportunities to cost-effectively increase the region's water supply through less environmentally destructive means, the EPA put a fork in Denver's dam.

This was the best news western environmentalists had heard since Echo Park Dam was cancelled in 1956. The EPA's veto of Two Forks marked a defining moment in water management in the American West. A leader of the opposition to Two Forks declared, "We have at this point seen the end of the big dam era."

But even a little dam can do enormous damage. Enter Windy Gap.

Chapter 8
WINDY GAP

Compared to the proposed Two Forks megadam that caused so much controversy, Windy Gap Dam is a pipsqueak of a structure. Yet the diminutive dam managed to generate major conflict.

Against a backdrop of rising public concern over water development, northern Front Range cities proposed the Windy Gap Project in 1967. Three counties—Boulder, Larimer, and Weld—were struggling with explosive population growth and anticipating even faster expansion in coming decades. Thirst is a potent motivator: When people are faced with the prospect of reservoirs emptying, their efforts to secure water can take on a kind of primal intensity.

The need to find new water supplies for crowded cities crashed into the environmental movement's momentum—which seems like the setup for an exciting showdown. But the conflicts that ensued involved piles of paperwork and endless courtroom proceedings: not exactly the stuff of high drama in the Wild West.

The Front Range needed water not for the farms and ranches of America's mythic imagination of the High Plains but for the banal suburbs that housed employees of companies like IBM and Hewlett-Packard, which began operations on Colorado's northern Front Range in the 1960s and '70s. As an urbanization boom driven by high-tech industry drew people to the area and gobbled up open space, homes and businesses replaced irrigated cropland. Potato patches were paved over with office parks. Fields of sugar beets gave way to subdivisions. And water once used for agriculture grew tract homes and shopping centers.

The flow from irrigation ditches that was transferred to urban use couldn't come close to quenching the thirst of the swelling cities. They spread toward each other in a megalopolis that seemed it might someday stretch in a continuous sprawl between Fort Collins and Pueblo. The Front Range's abundant sunshine attracted hordes of new settlers. But with much sun comes scant precipitation: The eastern plains are semi-arid, just one damp step above desert. When pioneers pushed westward in the late nineteenth century, conventional wisdom held that "rain follows the plow"—it was widely believed that changes to the land caused by homesteading would transform the West's climate, making it more humid and thus more habitable. In the second half of the twentieth century, a new wave of Westerners assumed technology would remove nature's constraints. "Rivers follow the suburbs" seemed to be their creed.

Citizens of the East Slope expected water to flow from the taps of freshly built subdivisions. They also expected West Slope rivers to provide them with rapids and trout and scenery. But you can't have your rivers and drink them too.

THE SITE OF THE proposed dam and reservoir that would divert more West Slope water to the dehydrated East Slope was a geologic feature known as Windy Gap. Located just below the confluence of the Colorado and Fraser Rivers a few miles west of the town of Granby, Windy Gap forms a slot in the mountains, funneling westerly winds. Gusts accelerate as they squeeze through this narrow passage and then rush downward toward the plains.

Windy Gap is a piggyback project: It uses the Colorado-Big Thompson Project's infrastructure to store water and send it to the East Slope. If the C-BT hadn't been financed and constructed by the federal government, the Windy Gap Project would not have been built. Aside from the physical gap in the mountains, another gap must be understood to make sense of the Windy Gap Project—the gap between how much water the C-BT was projected to provide and how much water, after the project was completed, it actually delivered.

The Bureau of Reclamation designed the C-BT to divert 310,000 acre-feet of water annually to the East Slope. But inaccurate hydrology studies and changes to the project's design during construction resulted in significantly less water being supplied: Something like 80,000 acre-feet less each year were delivered to the northeastern plains. That's more than a gap—it's a gaping hole. And the powerbrokers in northern Colorado were desperate to fill it.

The C-BT had resulted from the federal government's largesse, but this didn't stop Northern Water from complaining to the Bureau about the shortfall. The communities benefitting from the C-BT had paid pennies on the dollar for the project, but this didn't prevent them from feeling they were owed the full 310,000 acre-feet of water they had been promised by the government. Such is the nature of entitlements, whether paid in cash or in acre-feet of the most precious substance on the planet.

In response, the Bureau began studying ways to supplement the C-BT Project and fill the water supply gap. In 1952 the Bureau proposed increasing the amount of water the C-BT delivered by building a Windy Gap extension. A small dam on the Colorado River below its confluence with the Fraser River would corral the Fraser's spring floodwaters. Instead of being stored long-term behind the proposed dam, the captured water would be pumped uphill through a six-mile pipeline into

Lake Granby, the largest storage reservoir in the C-BT system. From there, the rerouted water would be sent with other C-BT water to the East Slope to help fill irrigation ditches.

But the Bureau decided the construction costs of the Windy Gap proposal were too high to justify a project that would provide more irrigation water to the plains of northeastern Colorado.

A decade later, however, desperate cities of the northern Front Range used the Bureau's Windy Gap proposal as a blueprint for a project to stave off the water impoverishment they faced due to soaring population growth.

The six cities pushing for the Windy Gap Project joined forces with Northern in 1970, forming the Municipal Subdistrict of Northern Colorado Water Conservancy District (MSNCWCD), adding another entity with an absurd acronym to a punishingly long list of complicated players in the upper Colorado River water drama.[10]

Environmentalists had also stepped onto the stage. Legal challenges from two fronts stalled the Windy Gap Project. Opposition by the West Slope, which viewed Windy Gap as yet another raid by the Front Range on its water, had been expected. Obstruction by environmentalists armed with loads of new federal laws had not.

The Subdistrict initially pursued funding for Windy Gap through the Department of Housing and Urban Development. When the red tape wrapped around that grant money proved too difficult to slash through, the Subdistrict figured out how to pay for the project itself by issuing bonds. Even though the federal government wasn't footing the bill for Windy Gap, it scrutinized the potential environmental impacts of the project and placed bureaucratic hurdles in the way, illustrating the U.S. government's shifting role in water development—from building dam projects to blocking them.

Resistance from federal agencies to Windy Gap Dam on environmental grounds foreshadowed the EPA's cancellation

---

10   To prevent brain damage, the MSNCWCD will be referred to as "the Subdistict."

of Two Forks Dam. Though Windy Gap is a less dramatic structure than Two Forks, and the controversy it spurred is less well known, Windy Gap was no less important a pivot on which the nation's water policy turned.

～～～～～～～

THE SUBDISTRICT'S FIRST ORDER of business was securing a contract with the Bureau of Reclamation to transport Windy Gap water through the C-BT. But environmentalists had been bombarding the Bureau with criticism of its water projects, so the agency stalled the contract by requiring an environmental impact statement (EIS). Few acronyms make the blood of dam boosters boil as reliably as EIS—and NEPA.

To make sense of the alphabet soup that prevented Windy Gap Dam from being swiftly authorized and built, we must flash back to 1969, the year Congress passed the National Environmental Policy Act, known as NEPA.

As the outrage of citizens over environmental degradation reverberated through Washington, Congress responded with NEPA. Signed into law by President Richard Nixon in 1970, this legislation directed the U.S. government to "use all practicable means...to create and maintain conditions in which man and nature can exist in productive harmony." Though this sounds like something said in Haight-Ashbury in the sixties, the details of NEPA were not hippie sentiment but the stuff of wonky science and tedious talk in courtrooms.

NEPA created a framework for the federal government to assess actions that could affect the environment, including dam construction. The potential impacts of a water project had to be addressed in an EIS, and alternatives to the project had to be proposed. NEPA required an EIS be made available for public use and comment throughout a project's review process, and the law charged the EPA with reviewing and commenting on any EIS that fell within the agency's responsibilities to protect the environment.

NEPA unleashed a flood of awkward acronyms, produced piles of paperwork that rose to heights reminiscent of the

Rockies, and created legal mazes as complicated as the canyonlands of the Southwest. It also focused scientific scrutiny on the upper Colorado River's degraded environment by inviting the public to weigh in on the pros and cons of the proposed Windy Gap diversion. Resistance and legal challenges from the West Slope and from environmental groups at every stage slowed the permitting process to a glacial creep and added considerable cost.

The West Slope questioned why the C-BT, a project with the stated primary purpose of providing supplemental water for irrigated agriculture, should be used to deliver water to sprawling urban and suburban development on the Front Range. This was an excellent question—and to this day it has not been adequately answered.

The West Slope also worried about the Windy Gap diversion adversely affecting "present or future uses" of water in the Colorado River Basin (which had been protected by Senate Document 80 and the state's Water Conservancy District Act of 1937). And the West Slope grumbled about the East Slope's wasteful use of water: Couldn't Front Range cities use water more efficiently and do more to conserve their supplies instead of raiding West Slope rivers? This question is still being asked—and will continue to be asked with increasing volume as the state's ongoing water wars intensify.

During the Windy Gap dispute, West Slope demographics had shifted since the days of statewide debate over the C-BT. Anger toward the East Slope, however, remained constant. Journalist Allen Best noted that when he lived along the Colorado River in Grand County in the 1970s and '80s, locals viewed diversions to the East Slope as "moral thievery and political thuggery." Opposition to water projects can make for strange bedfellows: At public meetings on the West Slope about transmountain diversions, dreadlocked hippies and gun-toting cowboys show up to voice their shared disapproval.

In 1974, the Subdistrict responded to the federal government's requirement of an EIS by sending environmental consulting firms to Windy Gap to assess the project's potential impacts. The firms looked around the site, and according to Gregory Silkensen's report *Windy Gap: Transmountain Water Diversion and the Environmental Movement,* they declared the impacts would be "quite minimal."

But anglers who leased fishing rights on private property near Windy Gap didn't agree. They worried about the project's impact on sections of stream they fished. And they wondered if a fish ladder should be built with the dam. People who spent time in the upper Colorado watershed near the proposed project believed further diversions would harm the river, but they didn't have scientific studies to support their concerns. Not yet, anyway. That would come much later—after it was too late to prevent widespread damage caused by the dam.

IN 1976, THE SUBDISTRICT's progress toward building Windy Gap was stymied when it learned that federal law required an EIS for the entire C-BT Project—spanning all the system's Continental Divide–straddling, mind-bending complexity.

The Subdistrict soldiered on. It delivered a comprehensive EIS to the Bureau. But the Bureau requested more detail. Concerns were raised over sewage released from the town of Granby seeping from the Fraser River into Lake Granby via the Windy Gap diversion. The Subdistrict submitted a rewritten EIS—it vanished in a Department of Interior vacuum in Washington, D.C.

To say that the proponents of Windy Gap were exasperated by the tedious permitting pace is to understate the matter. Daniel Tyler, in his book *The Last Water Hole in the West,* the preeminent history of the Colorado-Big Thompson and Windy Gap Projects, makes the case that Earl F. Phipps committed suicide in part because of his frustration with negotiations. Phipps served as manager of Northern Water when the Windy Gap proposal had to be run through a gauntlet of West Slope opposition and new federal environmental laws; this strain, Tyler asserts, contributed to Phipps ending his own life.

Five years after the EIS process had begun, the Subdistrict pressured the Bureau to release the draft EIS, which the Bureau finally did—all ten chapters and nearly three hundred pages of it. People who had complained about the tedium of reading Senate Document 80 had no idea how mind-numbing the paperwork piling up around transmountain diversions would become. The challenge of constructing the C-BT pales in comparison to the effort required to make sense of an EIS. With the implementation of NEPA and the EIS process, the public's attention span was strained, and environmental lawyers realized they would never want for work.

The EIS process invites citizens to participate in determining the fate of a proposed water project but then excludes them by demanding an extraordinary investment of time and effort to read a report. From the first dull sentence to the final mind-numbing paragraph, an EIS provides the intellectual equivalent of waterboarding. Those who survive the punishing process generally have a lot at stake—they stand to gain or lose something of great value if the proposed project is built. When a stretch of river you love may be harmed, you read the EIS, no matter how torturous, and then you show up at public meetings to make sure your comments are heard. This is the messy and difficult work of citizenship. It brings to mind Winston Churchill's statement: "It has been said that democracy is the worst form of government—except all the others that have been tried." The water development democracy created by NEPA is an awful way to govern a natural resource. But it's less awful than letting water developers kill the Colorado without the public having a say in how a river that sustains them is managed.

NEPA required the Subdistrict to respond to environmental concerns raised during a public hearing process. Before the final version of the Windy Gap EIS was completed, citizens at public meetings voiced their worries. Many of them called out the Subdistrict's claim that the water quantity delivered downstream from the proposed dam would be enough to maintain healthy fish habitat. Some of their concern centered on the Gold Medal trout fishery[11] on the Colorado River below the proposed dam site—Bud's future battleground.

Citizens also expressed concern over potential impacts to endangered species farther downstream—which brought into play yet another piece of legislation born of the burgeoning environmental movement.

The 1973 Endangered Species Act provided for the conservation of critical habitat that sustains threatened and endangered species. The legislation became a powerful weapon in the quiver of citizens struggling to curtail water development. When endangered fish species were identified in the Colorado downstream from Windy Gap, a new front opened in the battle over the proposed project.

The Colorado pikeminnow, despite its diminutive-sounding name, can grow to one hundred pounds or more. Once known as the squawfish, it was eaten by Indians and settlers and formed an important food source during the Great Depression. But as the nation got hooked on rainbow trout, Americans derided the squawfish as a "trash fish" and poisoned it to make way for species considered more fun to catch.[12]

Then the fortunes of the squawfish turned. It was renamed the pikeminnow for reasons of political correctness, and people started paying attention to its plight. Ecologists pointed out that the disappearance of a top predator like the pikeminnow can signal the collapse of an entire ecosystem—similar to the wolf that was eradicated from Yellowstone and then reintroduced to heal the park's damaged ecology.

Evolution equipped the pikeminnow and other species native to the Colorado River Basin like the humpback chub to persist in one of the world's most chaotic aquatic environments.

11   The Colorado Wildlife Commission designates streams that provide excellent opportunities to catch large trout Gold Medal Waters.

12   For a great read about how rainbow trout charmed humans into helping them expand their range and displace native fish species, see *An Entirely Synthetic Fish: How Rainbow Trout Beguiled America and Overran the World* by Anders Halverson.

These creatures thrived for millennia amid wild cycles of drought and flood. Natural cataclysms they could survive. Dams and diversions that disturbed the river's natural flow they could not.

Windy Gap would further deplete the Colorado, threatening endangered species downstream. Though the amount of water Windy Gap would divert was modest relative to other megaprojects already built in the Colorado River Basin, the cumulative impacts of multiple diversions were already pushing native species toward extinction. The nation's changing environmental values, combined with the timing of the Windy Gap proposal, forced the project to bear the full brunt of public pushback against water diversions in the Colorado River Basin. Likewise with salinity concerns.

The Colorado River picks up salts as it scours ancient seabeds. By depleting the Colorado's flows, water projects concentrate dissolved solids carried by the river. These projects also promote intensive irrigation in the Colorado River Basin's mineral-rich soils—causing salt to leach into the river through return flows of contaminated irrigation water. The less water that remains in the Colorado, the more salty it becomes: As the saying goes, "the solution to pollution is dilution."

Water delivered downstream in the shrunken Colorado had become so saline it was killing crops. By diverting more flow from the Colorado River headwaters, the Windy Gap Project threatened to make the salt problem worse—creating yet another environmental hurdle for the Subdistrict to surmount. In 1978, the Wilderness Society and Trout Unlimited joined the Environmental Defense Fund in filing a lawsuit based on salinity concerns against the federal government. Salt and endangered species nagged Windy Gap—along with the West Slope's worries that the project would impede its water interests.

Before concrete could be poured for Windy Gap Dam and its pump plant, the project's proponents had to wage a war of attrition on several fronts. Savvy attorneys and vast sums of money, along with unyielding persistence, were required to prevail.

The Subdistrict survived the herculean trials of placating the West Slope, surmounting each environmental hurdle, and securing all twenty-three permits and licenses required to move the project forward. In 1981, construction of Windy Gap Dam began. In 1985, nearly two decades after this relatively small project had been proposed, the dam was finally completed. And the river's ancient voice was silenced as flowing water stalled behind the wall.

Several years later Bud would have to muster significant persistence of his own, along with considerable financial resources, when he began his battle to undo the damage wrought by Windy Gap Dam.

WINDY GAP RESERVOIR WAS designed to pool water to be pumped to Lake Granby. Instead of serving as a deep storage reservoir, Windy Gap forms a shallow collection basin that stalls the river's flow prior to pumping. And it collects more than water.

What's astounding about Windy Gap is that for all the pitched battles over the proposed project, the actual damage this shallow reservoir caused the river was not anticipated. The law of unintended consequences surprised both engineers and environmentalists, providing a sobering lesson for everyone. To paraphrase the biologist J. B. S. Haldane: Nature is not only more complex than we imagine, it is more complex than we *can* imagine.

The first problem to emerge after completion of the Windy Gap Project was a pathogen that spread like a plague through the dammed waters.

Part III

Chapter 9
THE WHIRLING DISEASE
NIGHTMARE

In the state of Colorado, whirling disease has devastated many rivers, but few have suffered as much as the Colorado River below Windy Gap. Before whirling disease infected this reach, the Colorado supported one of the world's most productive wild rainbow trout fisheries.

The rainbows that thrived in the upper Colorado River reproduced naturally—biologists didn't need to support the population with stocking. Nature created abundance, and the robust trout in these storied waters provided some of the best broodstock in the state for twenty-five years. Biologists from the Colorado Division of Wildlife electroshocked the river during spawning season to temporarily stun the fish. Then they squeezed streams of milt from males and vibrant orange eggs from females. In hatcheries they reared Colorado River rainbow fry, which they used to replenish rivers across the state with hardy, long-lived trout—the pride of the Colorado fisheries system. But whirling disease wiped out most of the wild rainbow trout downstream of Windy Gap, erasing the genetic bounty contained in these "super trout."

Whirling disease is caused by a microscopic parasite (*Myxobolus cerebralis*) that attacks the neurological and skeletal systems of young trout, causing them to whirl in a weird circular pattern—which makes feeding difficult and turns infected fish into easy prey. Black tails, deformed bodies, and disfigured heads with bulging eyes are telltale signs of the disease. First detected in rainbow trout in Germany in 1893, whirling disease has caused trout fisheries in the United States to crash. The disease illustrates how a pandemic can circle the globe and decimate a population: Whirling disease is the smallpox of the fish world, the Ebola of waters where young trout dwell.

While rainbow trout were disappearing from the Colorado River below Windy Gap, Bud and other anglers noticed the brown trout population was exploding, filling the void created by the missing rainbows. Brown trout fare better in waters infested with whirling disease because they evolved in Europe and western Asia, where whirling disease originated. Through many generations of exposure to the parasite, natural selection provided the brown trout species with some resistance. Rainbow trout, in contrast, evolved along the Pacific Rim, their range spanning the West Coast of North America and the Kamchatka Peninsula in eastern Russia—far from where the whirling disease parasite evolved. Because rainbow trout were exposed to the parasite more recently than brown trout, the rainbow species hasn't had time to develop resistance.

The rainbow's susceptibility to whirling disease reminds Bud of Native Americans' lack of immunity to smallpox; because they had no exposure to the pathogen before Europeans spread it to the New World, the consequences were horrific. Bud sees whirling disease as a harbinger of what lies on the horizon for humans in our globalized age, when people across the world stricken with virulent diseases are a plane ride away from spreading a pandemic across the planet.

The whirling disease parasite in all its gruesome complexity seems like something conjured in the mind of a sadistic madman obsessed with Rube Goldberg machines. The parasite spends part of its convoluted lifecycle inside a species of small aquatic worm, *Tubifex tubifex*, a relative of the earthworm. Tubifex worms ingest myxospores, a form of whirling disease parasite spore round and tiny—roughly the size of a blood cell. A myxospore opens inside the worm's gut, morphing into a second form of spore: a triactinomyxon, or TAM for short.

Much larger than a myxospore, a TAM is shaped like a three-armed grappling hook. After a TAM is released from an infected worm, it floats in the water until latching on to an unlucky trout. The TAM pierces the fish's skin and injects cells of the parasite.[13] Once inside, the parasite travels through the trout's nervous system, damaging the organs of equilibrium— sending young fish swimming in senseless circles.

The parasite multiplies as it devours cartilage, deforming a young fish's developing skeletal system.[14] While inside a trout, the shape-shifting parasite morphs back into a tiny myxospore. When the fish dies, its skeleton decomposes, releasing myxospores to settle in the stream bottom, where tubifex worms feed in the mud.[15] And so the process continues, in a cycle ghastly for the fish and glorious for the parasite, putting a sharp point on an unsettling truth: Nature can be cruel to the core. Humans have a rare ability in the animal kingdom to show mercy to other species—and we have the unusual capacity to serve as stewards of our surroundings instead of devouring everything in sight. Saving the Colorado River, rather than devouring its waters to support cancerous growth, is a crucial test of our humanity.

SCIENTISTS WHO STUDIED THE upper Colorado River realized tubifex worms were abundant in Windy Gap. Because the reservoir had been designed not as a deep storage basin but as a

---

13   A TAM can also enter a trout if the fish feeds on an infected tubifex worm.

14   Older fish are less susceptible because much of their cartilage has already hardened into bone.

15   Alternatively, an infected fish is eaten by another trout and passes through its digestive tract, releasing the myxospores into the unlucky predator that ate this contaminated meal.

place to pool water for pumping, the dam formed a shallow impoundment filled with silt. To this muddy brew were added organic nutrients from cow pies of ranches upstream, and from waterfowl that deposited their droppings in the nutrient-rich soup. Worms in the muddy reservoir that became infected with the whirling disease parasite released TAM spores. The TAMs then drifted in dense plumes downstream of the dam to infect trout, laying waste to the fishery.

When Windy Gap Reservoir became a breeding ground for the whirling disease parasite, the rainbow trout population in the river downstream crashed. From some five thousand rainbow trout per mile in 1981, the count fell into the hundreds in the 1990s—and hovered near zero in 2010. But those are just numbers.

What an angler saw in 1981 were silver snouts of rainbows poking above the river's surface as the fish sipped insects. So many rings spread across the water from feeding rainbows you couldn't count the rises. Rods doubled over from the weight of the fish, and reels screeched with metallic zings as line zipped from spools when giant trout surged and leapt in brawny struggle. Twenty years later, a rainbow of any size tightening an angler's line was so rare it merited discussion.

WHEN FISHERIES BIOLOGIST Barry Nehring talks about his career, he often refers to "the whirling disease nightmare."

So packed with coiled energy he seems larger than his medium build, Barry tells me he became a field biologist because he couldn't stand working inside. He has now spent four decades walking the banks and wading the waters of most major rivers in the state of Colorado, as well as countless small streams and high-country lakes. Like Bud and others who spend a lot of time on the water, Barry noticed rainbow trout disappearing from the state's rivers in the early 1990s. "Fish were dying like flies," he says. After uttering this sentence, Barry criticizes himself for using "flowery language." He tells me scientists shouldn't let their emotions lead them to overstatement. But I've found that Barry's passion for wild trout fills him with a verbal enthusiasm he cannot contain. "Put a nickel in him and you'll get five dollars' worth," notes Bud, no stranger himself to colorful phrasing.

The first time I mention whirling disease to Barry, thirty minutes later he's still talking. How he can breathe while machine-gunning his words without pause is a mystery. Maybe his extraordinary lung capacity, revealed when we hike together, accounts for his tireless ability to talk. What Barry likes to talk about most is trout. The information he has accumulated through many years of reading scientific reports and conducting field research has turned his brain into a priceless archive—a Smithsonian of trout knowledge.

In his early seventies, Barry is officially retired as a fisheries research biologist employed by Colorado, but he occasionally works for the state and continues to publish peer-reviewed scientific papers. At Colorado State University in Fort Collins, he began his studies in 1970 under Robert Behnke, a legend among scientists who study salmonids (the family of fish that includes salmon, trout, and related species).

Dr. Behnke, who died in 2013, was recognized throughout the world for his expertise in the classification and conservation of native trout species. Known as "Dr. Trout," Behnke was also a great raconteur. Both his scientific aptitude and his storytelling skill have been passed on to Barry, whose passion for studying trout and talking about them is so intense he seems on the verge of spontaneously combusting when spinning tales about his mentor, or speaking about the damage whirling disease has wrought in Colorado's rivers and lakes.

Whirling disease was introduced into the United States in the 1950s when Denmark created a frozen rainbow trout industry, Barry explains. The fish were raised in Denmark; frozen fillets were exported to the United States. Because the fillets didn't sell well in supermarkets, they were ground up and fed to fish in Pennsylvania's hatcheries, which then tested positive for whirling disease.

Barry also shares this story: In June 1965, a Danish freighter with a damaged freezer docked on the Pacific Coast near

Monterey, California. Rainbow trout fillets that had thawed were useless as food for the crew. The owner of a private fish hatchery ground up the fillets and fed them to trout in his ponds, which discharged water into a nearby creek. In January 1966, whirling disease was confirmed in this private hatchery. From there, presumably, it spread into local waterways.

When Barry and his colleagues conducted electroshock studies of the upper Colorado River in the early 1990s, they noticed a glaring absence of young rainbows. "It was like aliens had come in and vacuumed up the rainbows and left the browns," Barry says. In all his years of working on rivers, he'd never seen anything like this. He knew something was wrong. Very wrong.

As young rainbows continued to disappear but adults remained in the river, Barry suspected whirling disease was the culprit. But at this point the disease was poorly understood. As with other parasites that require two hosts, whirling disease presents researchers with plenty of problems to solve. The stages of the parasite's lifecycle, the biology of its host worms, why the disease hadn't devastated trout populations in other states: These were all unknowns.

Equipped with empiricism and logic, Barry set out to solve the mysteries of whirling disease. When he found that all rainbow fry had vanished in a stretch of the Colorado save one creature with bulging eyes and bent spine, he sent this disfigured specimen to a lab. It tested positive for whirling disease. Barry reasoned that other rainbow fry had been killed by the parasite and had decomposed in the river.

To add evidence to his hypothesis that whirling disease was causing this catastrophic die-off, Barry put live fish inside cages in rivers. Caged fish in the Colorado died in droves; when autopsied they tested positive for whirling disease.

Big rainbows in the river were still bending the rods of anglers. But the three youngest age-groups had vanished. When the adult rainbows died off, there wouldn't be new generations to replace them. Barry saw disaster on the horizon. In December 1993 he wrote a white paper for the Colorado Division of Wildlife suggesting whirling disease was taking such a heavy toll on Colorado River rainbow trout it could send them "the way of the passenger pigeon."

Reactions to Barry's whirling disease hypothesis ranged from skepticism to anger. He was accused of spreading fear over something that didn't pose a problem. Whirling disease had been in the United States for several decades, but no widespread negative effects had been reported in the wild. Outbreaks had infected fish in hatcheries, yet rivers continued to support healthy populations of rainbow trout in states where whirling disease had been detected. Officials in Colorado insisted there was no need for panic.

Stocking fish is big business: Some nine million rainbows a year are reared and released into Colorado's lakes, rivers, and streams. If the state stopped stocking, fewer fishing licenses would be sold, meaning less revenue for the Division of Wildlife. A lot of dollars from anglers were at stake, as was a lot of money invested in hatchery infrastructure. Shutting down the state's hatcheries that tested positive for the disease was not a popular position.

Other states had panicked when whirling disease showed up in their hatcheries. In the 1960s and '70s, California destroyed millions of rainbow trout and closed down multimillion-dollar hatcheries to control the parasite, Barry explains. Michigan dumped chlorine into a river to disinfect it; other states killed and buried infected fish. These extreme measures seemed unwarranted to administrators of Colorado's fisheries. They decided to continue stocking the state's waters with trout from infected hatcheries—even though Barry insisted this was a terrible idea. He argued stocking programs should be managed to stop the whirling disease pathogen from spreading through rivers and streams.

Barry was convinced catastrophe was imminent. He was not alone. Fly fishers like Bud and Charlie Meyers, a renowned outdoor writer for *The Denver Post,* also raised warning flags. These citizen scientists reported crashing rainbow trout populations in the rivers they frequented. Biologists like Barry bolstered these observations with scientific studies that documented dramatic declines in fish numbers. But scientists and reporters who worried publicly that whirling disease was impacting wild trout were derided as overreacting doomsayers.

The divide between field biologists in Colorado who gather data and administrators in Denver who develop policy can seem as deep as the chasm carved by the Gunnison River in Barry's backyard. Yet Barry's rigorous studies and his determination to communicate his findings managed to cross this divide. Resistance to the idea that whirling disease was decimating the state's wild rainbow trout waned when the results of Barry's research combined with news from Montana that its world-famous rainbow fishery in the Madison was crashing.

Whirling disease, a problem that a few years earlier a handful of people had heard of, suddenly became a topic of frantic discussion and debate among anglers and fisheries managers. For his early sounding of the alarm, Barry went from being a pariah within the Division of Wildlife to employee of the year.

"Seeing was believing," says Barry. Early in the disease outbreak, he walked to a riverbank where he'd taught one of his sons to fly-fish, and he watched a young rainbow whirl out of the water and thump onto the bank with its crooked spine contorted. He says, "I nearly cried when I watched the fish corkscrew its way out of the stream."

When trout are frightened they dart for cover. Barry explains that when a whirling disease–infected fish with a warped spinal column and a damaged brainstem experiences a fright response, it chases its tail like a deranged dog: It twists and spins so strangely and violently it sometimes shoots itself out of the water when scared. Barry says at the height of the whirling disease epidemic in the Colorado River, when he stomped on a bank, rainbow fingerlings would spin into the air "like raindrops falling in reverse."

When Ronald Hedrick, an epidemiologist at the University of California, Davis, came to Colorado for a week to see what was going on, Barry took him to the Colorado River and showed him fish with black tails, a telltale sign of whirling disease. As the men moved upstream toward Windy Gap to sample fish, 90 to 95 percent of the brown trout had black tails and bent spines. Hedrick, one of the world's top authorities on whirling disease, took some disfigured fish back to his lab to study. He reported to Barry he'd never seen such cellular damage. "He became a believer," Barry says. There was no denying that whirling disease was wiping out fish in the Colorado River below Windy Gap.

Acceptance that the whirling disease pathogen was affecting Colorado's rivers turned into concern. And concern shaded into panic. Without trout in its rivers and lakes, the state would be a much diminished place, and its multibillion-dollar tourism industry would be eviscerated.

In late 1994, as whirling disease worries intensified across the West, one afternoon Barry answered the phone. A man named Karl Johnson asked Barry if he remembered him. Barry recalled Dr. Johnson from twenty-two years previous, when the men had played bridge together at the Ginger Quill, a fly-fishing club on the North Platte River. Karl Johnson was one of many club members who were accomplished virologists, pathologists, and immunologists. Their sharp minds had left a strong impression on Barry.

Karl Johnson made a name for himself as one of the discoverers of the Ebola virus and Hantavirus, and he was depicted in the book *The Hot Zone* as the director of the Centers for Disease Control laboratory in Reston, Virginia—where Ebola, one of the most lethal diseases known, nearly escaped the laboratory. Horror writer Stephen King, master of the macabre, called the factually accurate book "one of the most horrifying things I've ever read."

After whirling disease was discovered in Montana's Madison River in the fall of 1994, Karl, an avid fly fisherman, had spent a week digging through the scientific literature on whirling disease at Montana State University. He told Barry, "I'm the co-discoverer of *Ebola zaire* and this parasite scares the hell out of me. Tell me about whirling disease in Colorado."

"It's a nasty story," Barry replied. And he didn't pause for breath until he'd filled Karl in on every detail.

Karl became so concerned about the pathogen, he rounded up a group of likeminded people and founded the Whirling Disease Foundation, a nonprofit based in Bozeman, Montana, that raised several million dollars to fund whirling disease research.

No other state relies on its trout fishing industry as much as Montana. When young rainbows vanished from the Madison, one of the West's most celebrated fly-fishing rivers, many people switched into crisis mode. Scientists put aside their egos to work together and focus on the common threat, Barry explains. He says the endeavor was a fine example of the way science should be done—no politics, no power struggles. The nightmare of whirling disease was cause for the scientists, most of whom were fly fishers, to pool their resources so they could defeat the disease that threatened the rivers they loved.

IN THE AUTUMN OF 1997, Barry got a call from Bud Isaacs. Bud was asking for help with the whirling disease disaster on his stretch of river. Barry says he had fielded strange phone calls ever since news of the disease had broken, and weird stories increased at the height of whirling disease hysteria. Someone called him to share a theory that whirling disease had been introduced to the state in anchovies for fish food. Another person phoned him to insist the parasite had been in Colorado forever. One man was convinced his wife had whirling disease.

So Barry wasn't sure what to make of Bud at first—Barry says he "sort of blew him off." But Bud persisted, calling Barry back to explain that below Windy Gap he'd seen the rainbows disappear. He also said he could find funding for Barry to study whirling disease in Windy Gap Reservoir. Barry, who had started referring to Windy Gap as "a biological Chernobyl," was intrigued. He had been studying the numbers of whirling disease spores above and below Windy Gap—TAM spores skyrocketed in the river reach below the reservoir.

The word "reservoir" invokes images of a waterbody deep and clear, but the water that backs up behind Windy Gap Dam Barry calls a "putrid collection pond." This shallow "pond" is over-rich in nutrients from the waterfowl crap that falls in it and the cow shit that washes in from surrounding fields.

Barry suspected the reservoir was an incubator for the disease; he reasoned Windy Gap supported a profusion of the tubifex worms the parasite needs to complete its lifecycle. He hypothesized the parasite would be most abundant in a quiet corner of the reservoir, where stagnant water and mud thick with organic matter would support many worms. If he could figure out where the parasite was concentrated in the reservoir, he could develop a management plan to sequester it, preventing spores from spreading into the river downstream. But to test his ideas he needed funding.

Bud claimed he could raise the money for Barry's research over coffee in an afternoon. And that's exactly what he did.

Bud found the money needed for research; then he let Barry do the best science possible so he could figure out what was going on in Windy Gap Reservoir. "There were no strings attached," says Barry. He insists neither Bud nor anyone else at Chimney Rock Ranch who helped fund his whirling disease studies pressured him to change his data or his conclusions. He says he wishes science could always be done that way—without persuasion from politicians, without manipulation from interest groups.

Science is seldom so pure. Researchers' findings are often secondary to the agendas of funders. Barry points out that Bud's stance of letting scientists do their work unimpeded has, in his four decades of experience in fisheries biology, served as the exception rather than the rule. He sees Bud as a vigilant watcher of the river and an unapologetic whistleblower. "He's willing to stick his neck out," says Barry, whose strong voice breaks when he speaks of Bud's commitment to studying and saving the Colorado.

On the steep canyon walls cleaved by the Gunnison near Barry's home in Montrose, he can out-hike most twenty-year-olds. He carries packs loaded with a hundred-plus pounds of meat from deer he shot. While working with electricity in rivers he has been shocked and nearly drowned. But Barry's leathery toughness conceals a tender heart.

Barry once fed a starving mountain lion by dangling a sandwich from a boat paddle toward the emaciated creature.

He and his wife have adopted six kids and fostered some forty more, along with raising three of their own. When Barry reads to his many children "The Worth of a Wild Trout," the transcript of a presentation from a wild trout symposium, he rarely gets through it without choking up.

As a scientist, Barry knows that trout live only in clean water—they can't tolerate contamination. The temperature, turbidity, and morphology of a river must be pristine for wild trout to thrive. All that is lovely in this life is reflected in the smooth skin of a trout as it slides through a mountain stream, its sides flashing silver as it turns, its streamlined shape vanishing in deep pools. A trout rises like an emotion at the water's surface and then sinks back toward hidden depths, and we watch without breathing, our hearts expectant and full. To know trout are there, in clean water, moving between sun and shadow as they pass through unknowable currents, lifts us in ways that exceed the economic value a productive fishery provides.

Barry says, "As long as there are people like Bud who care about wild trout, the rivers will be okay." His eyes glisten as he tells me this. To Barry, trout are more than fish: Trout are critical indicators of a river's health. And healthy rivers contain more than moving water. Through their channels the beauty of the world flows.

AFTER BUD SECURED FUNDING for Windy Gap research, Barry took samples from the reservoir's muddy bottom to analyze worms. The results showed his hypothesis about the location of a hot zone was wrong. The worms the whirling disease parasite needs to complete its lifecycle were clustered near the inflow where the river enters the reservoir—not, as Barry had suspected, in a stagnant corner.

A good scientist constantly scrutinizes facts and questions his own conclusions. Barry analyzed the results of his study and then set out to explain the facts with a new hypothesis. His revised thinking was that the river's current carried myxospores from infected fish into Windy Gap Reservoir; these myxospores settled out of the flowing water as it collided with the slack water of the reservoir. In this concentrated zone, worms became infected and released TAM spores, which drifted from the reservoir to infect fish downstream. That's why the rainbows had disappeared below Windy Gap Dam, Barry concluded.

Barry reasoned that the entire whirling disease incubator of Windy Gap Reservoir needed to be separated from the Colorado. A bypass channel to reconnect the river's flow around the dam would prevent the muddy, worm-ridden "putrefaction pool," as Barry calls the reservoir, from producing spores that drift downstream in dense clouds, decimating wild trout.

And why are wild trout so important? Why not just refill a waterway with hatchery fish when disease wipes out a wild population? What is lost when wild trout disappear from a river?

Trout that live and breed in streams must survive in varied environments and adapt to changing conditions. Differences in behaviors and physical traits that allow some fish to thrive in the wild arise from genetic diversity. This diversity is lost in hatcheries, where fish aren't forced to struggle for their survival. Trout churned out by hatcheries are cookie-cutter copies that lack the vigor and sheen of trout that breed in the wild. Inbred fish that gorge on food pellets in hatchery raceways are to their wild cousins what a domestic turkey is to its counterpart in the forest—a gorgeous, savvy creature bred to dimwitted blandness in captivity.

A core concept of conservation biology is that genetic diversity is essential for animal populations to survive in the wild. Maintaining a robust and varied gene pool promotes the long-term survival of a species in the face of threats such as disease outbreaks and climate change. But wild trout also possess an intangible worth not reflected in the mathematical models of scientists.

What is the worth of a wild trout? Those who spend their lives pursing them might ask, as Reverend Dan Abrams did in his speech that Barry reads often to his children, "How do you slap a price tag on a dream?"

Fish live in Windy Gap Reservoir. You can watch pelicans herding them as the birds tip their tails in the air and dunk their heads in unison. The spectacle is so fascinating I sometimes forget I'm surrounded by paved parking places and concrete pathways. As I stare at the pelicans in the distant water, the chain-link fencing and barbwire that guards the waterworks blurs and almost disappears. Sometimes when I grow tired of watching birds through the security fence around the pump plant, I wander toward a gate with signs warning people away, and I tempt the employees of Northern to glare at me for getting too close. This is wildlife viewing in the same way that Guantanamo Bay is a vacation resort.

True wildness isn't partitioned from the public by security fences. Fences surround not only Windy Gap Reservoir, but also hatcheries, where fish are farmed like genetically modified corn in fertilizer-doused fields planted and harvested by machinery.

The palette that paints a wild fish cannot be reproduced in a hatchery. Only natural processes unimpeded by human intervention can create something so exquisitely sculpted, so vibrantly colored. Like a striking image in a dream, the sight of a wild trout swimming through a stream stirs us in ways difficult to describe.

What we know for certain is that the rivers and the land and the sky to which these fish are connected must be protected. The places that shelter these bright dreams must be safeguarded so our children can watch a wild trout rising from the deep pools of a stream. Barry and Bud have understood this for as long as they can remember. So has Tony Kay, who joined them in their quest to restore the upper Colorado.

Chapter 10

THE WORTH OF WILD TROUT

When Tony Kay arrived in Colorado in the late 1970s from his hometown of Johannesburg, South Africa, he had no job, no prospects. He found work assisting the bookkeeper of a pizza manufacturer and then landed a position with a management consulting firm, where he got in the habit of looking over the shoulder of a programmer and offering advice about code.

Impressed with Tony's aptitude, the programmer encouraged Tony to learn the language of computers and apply it to solving real-world problems. Tony knew nothing about computers, so he hesitated. But he mustered the confidence to teach himself programming. And in short order he had started his own company and was developing a land-lease system for an oil and gas bureau, learning as he went, growing his business with each problem he solved by building lines of code.

As reserved as Barry is effusive, Tony tells me in a quiet, steady voice that his software engineering business became a success in large part because he approached problems by working backwards. He would identify a desired outcome and then methodically reverse-engineer a solution. He applied this technique when he led the Colorado chapter of Trout Unlimited in the battle to get rid of Windy Gap Dam.

* * *

ASKING TONY IF HE has always been an outdoorsman is like asking someone if he has always breathed oxygen. A passion for pursuing fish and game in wild places is as much a part of who he is as his aptitude for solving problems. Bud calls Tony's knowledge of insects "unparalleled" and says that when Tony is pursuing birds he is always listening. "He knows every birdsong," says Bud. "And when we're turkey hunting, he can call in gobblers better than anyone I've ever known. He can hear gobblers in the distance when the wind is blowing snow and no one else can hear a thing. He's a fly tier, a model builder, and he plays guitar in a band. He's very different from my American friends."

Tony says he loves living in Denver because the city has easy access to the outdoors. And he loves living in the United States because of the abundance of public lands. He notes that the amount of land where people can hunt and fish in South Africa is small compared to the bounty of wilderness available to Americans. Soon after arriving in Colorado, he joined Trout Unlimited, which works "to keep our country's coldwater fisheries and their watersheds safe from environmental threats for this and future generations of anglers to enjoy." According to Chris Wood, president of TU, the organization was founded in 1959 "by anglers frustrated over the state of Michigan masking habitat loss by pumping out ever-more hatchery trout rather than repairing damaged rivers and streams."

Largely because of TU's pivotal role in the Two Forks victory, the organization became a force in Colorado powerful enough to tackle the damage caused by Windy Gap Dam. Tony explains that back in the late 1970s, when the permitting process for Windy Gap was underway, TU was concerned about the proposed project's impacts on the Colorado River.

But TU's leaders sensed a big fight brewing over Two Forks. Because TU was a small organization with limited resources, it had to choose its battles—it decided to focus on Two Forks.

In 1980 TU had 14,000 members nationally; now it claims nearly that many in Colorado alone. Tony explains that one factor that made the David of TU so effective against the Goliath of Denver Water was a sophisticated public relations campaign. The Colorado chapter of TU hired independent PR firms to help craft their message about what was at stake: the devastation of a Gold Medal fishery in Cheesman Canyon. And TU joined with Audubon and other environmental organizations in a grassroots effort that made inroads with the press.

At the first Save the South Platte Day in 1986, TV and radio stations showed up, and the event garnered front-page coverage in both Denver newspapers. Tony explains that Mark Obmascik, Pulitzer Prize–winning journalist and author of the birding classic *The Big Year*, wrote a two-page spread in *The Denver Post* on Two Forks that generated public support for the efforts of TU and other organizations opposing the project. Burying a free-flowing river teeming with trout beneath the dead water of a reservoir did not sit well with many citizens. A popular bumper sticker declared "Dam the Denver Water Board."

Tony says, "TU made sure nothing was neglected." When opportunities arose for the organization to get its message into the press, its leaders made sure their story was told. When public meetings about the dam were scheduled, TU sent representatives to speak. And TU lobbied relentlessly, building support with key players, ranging from the Colorado Division of Wildlife to the EPA in Washington, D.C.

Tony explains that Denver Water had a strong legal team. Because the organization was overconfident it would prevail in the courts, it didn't try to make its case to the public. Though Denver and some forty suburban allies ponied up $40 million for an environmental impact statement (EIS) that led the Army Corps of Engineers to approve Two Forks, they invested little in selling their project to Denverites, many of whom were anglers and birders who chose to live in the city and its suburbs because of proximity to natural gems like Cheesman Canyon.

In *A Ditch in Time*, historian Patty Limerick explains that because Denver Water had become complacent, the organization didn't invest enough resources in delivering its message about why the dam needed to be built—which allowed opponents to trounce Denver Water in the arena of public opinion. Anger over the damage the dam would cause Cheesman Canyon was heard loud and clear by the EPA, which had the final say on whether the project could proceed.

Tony explains that a key factor in the dam's defeat was savvy opponents "took away the specter of tree-hugging, wild-eyed environmentalists" by proposing sober plans to provide the city of Denver and its suburbs with additional water supply. Most notably, hydrologist Daniel Luecke, a former Notre Dame quarterback, led the Environmental Defense Fund in creating a comprehensive plan that emphasized water conservation, efficient use of existing supplies, and building small projects to provide the same amount of water that Two Forks would provide—but at a lower cost, and with less damage to the environment. Luecke and his collaborators used facts and data to challenge Denver Water's assumptions about population growth and future water requirements. Tony says, "We had guys who did their homework."

William Reilly was the EPA's national administrator under President George H. W. Bush, who had run on the promise of being "the environmental president." Reilly concluded that too many opportunities for water conservation and making use of alternative water supplies existed to justify the "very heavy, final and irremediable loss of an environmental treasure of national significance." His veto put a stop to Two Forks Dam, marking a decisive victory for TU—and signaling a change in how the environmental movement could successfully defend embattled rivers.

Tony says, "The real visionaries in the process of preserving natural resources are people like Dan Luecke who are prepared to look outside the box and find creative solutions to fixing problems that are totally in synch with the ideas and the aspirations of both parties."

The pragmatic campaign that led to the defeat of Two Forks contrasts to the idealistic fight against Echo Park Dam. Opponents of Echo Park said no to the dam in Dinosaur National Monument under the premise that national parks and national monuments should be pristine wilderness unsullied by water projects. Opponents of Two Forks said yes to increasing Denver's water supply—but without building the costly dam.

Tony advocates creating solutions to environmental conflicts that provide "both parties with what they need." He says, "Water buffalos and environmentalists tend to suffer from tunnel vision—they refuse to see their opponent's point of view." The success of the Two Forks fight confirmed his belief that for environmentalists to be successful in their opposition to dams, they must develop science-based alternatives that take into account the aims of their adversaries. Tony is driven by a desire to preserve healthy rivers. But, like Daniel Luecke, he uses the tools of science and technology to define the problem and develop a solution that works for everyone.

After Two Forks Dam was vetoed in large part because of TU's efforts, the organization was no longer viewed as a fringe group. Tony says, "The victory gave TU a seat at the table." When whirling disease hit Colorado in the early 1990s, under Tony's leadership TU leveraged its mainstream status to take the state of Colorado to task for its handling of the crisis.

While fisheries officials insisted that whirling disease posed little threat to the state's wild trout, TU raised alarms. Its leadership paid close attention to Barry Nehring's reports of disfigured fingerling trout doing the whirling dance in the state's rivers and streams. They cited Barry's studies of caged fish in rivers dying, and they noted the troubling decline in rainbow trout populations observed by TU members.

When Tony became president of Colorado Trout Unlimited in 1995, he decided the Division of Wildlife (DOW) was not taking seriously the threat posed by whirling disease. Even though hatcheries tested positive, fisheries officials refused to take steps to get rid of whirling disease, and they didn't recognize that the pathogen could spread from hatchery fish to wild trout. Tony acknowledges the DOW's reluctance to act was driven in part by a shortage of information—a scientific picture was still developing of the parasite's complicated life-cycle, how it spreads through free-flowing waters, and how it affects wild fish. But lack of motivation to address the crisis also stemmed from the culture of the state fisheries system, which placed high value on stocked fish and little worth in wild trout.

Public reservoirs stocked with hatchery trout were big business. At the state's flatwater moneymakers, anglers could keep heavy stringers of trout they caught. Colorado's efficient hatchery system replenished the trout supply, and hatcheries were rewarded with revenue from slews of fishing licenses sold to tourists and residents who bought Styrofoam buckets of worms and filled their frying pans with fresh fillets. Catching and releasing fish in remote river reaches generated little revenue, and fly-fishing for wild trout was seen as a bit elitist by the powers that developed fisheries policies.

Trout Unlimited made the case that Gold Medal fisheries for wild trout could drive truckloads of tourist dollars to Colorado, as in Montana. Through tireless effort, TU convinced the state to implement catch-and-release programs. But even with some catch-and-release regulations in place to support wild fish in a few rivers, through the 1980s Colorado still heavily favored hatchery programs over wild fisheries. Like assembly lines, hatcheries churned out trout that anglers could catch and keep. Protecting wild trout—with their superior genetic fitness, with their vivid colors and the crisp lines of their fins not rounded down from swimming circles in concrete tanks—was not a priority for the state of Colorado.

Late into the 1990s, state officials maintained there was no evidence that whirling disease could spread from hatchery fish to wild fish in open waters outside hatcheries—even though Barry's research indicated otherwise. But TU heeded Barry's early warning that the state was on track to lose nearly all its wild trout within ten years. Tony says, "We were informed by people like Barry that the shit was going to hit the fan."

The executive committee of Colorado TU went into crisis mode. They met with national TU representatives, who announced they would file suit against the Colorado Division of Wildlife for violating the Clean Water Act by knowingly discharging a biological contaminant into the state's rivers and streams. Tony asked them for a chance to turn the DOW around. Trout Unlimited granted him some time before they filed suit.

Eddie Kochman, DOW's aquatics manager, had a reputation for inciting constitutionally calm people to fits of rage. Tony, a consummate negotiator, seems as centered and serene as a monk. He went to meet with fiery Eddie and calmly issued an ultimatum: The DOW could either be sued by national TU—and it would lose, and the DOW would look bad—or it could impose a moratorium on all stocking until a thorough scientific analysis of whirling disease could be completed. According to Tony, Eddie said he needed to consult with the head of the division. Tony stood firm and insisted the moratorium be implemented right away. "And Eddie did it," Tony says. "He imposed the moratorium, even though it wasn't clear if he had the authority to do so. This was a courageous move on Eddie's part."

Whirling Disease in the United States, a report issued by TU in 1996, was the first comprehensive assessment of scientific knowledge and management practices related to whirling disease. Peer-reviewed by a panel of experts, the report proposed a blueprint for future research projects and management practices to minimize damage caused by the pathogen to wild trout. But producing sound science was only part of the battle. Trout Unlimited had to blast through the institutional barriers that had turned trout management in the state of Colorado into an enterprise rather like industrial livestock production— units were cranked out in tremendous quantity at the expense of quality.

To the DOW, each stocked trout in the state's waters was a dollar sign. When the moratorium began, many dollars disappeared. Powerful interests throughout the state shouted for an immediate resumption of stocking. Colorado decided to put together a whirling disease commission composed of diverse stakeholders. Tony explains that all stakeholders except TU favored continued stocking. Trout Unlimited was a single outlier advocating for wild trout. Tony knew he was outgunned and needed to find common ground with his adversaries.

He asked every group on the commission to tell their story and then listened closely. They all wanted to catch trout, he realized, and they all wanted the trout to be healthy. Tony and his team went to work creating a policy the DOW could institute post-moratorium—and to which all stakeholders in the whirling disease commission could agree.

Tony says, "TU did their homework." They also did the work of others on the commission for them so everyone could go along with the plan that TU proposed. The key was establishing zones with different stocking regulations. For example, to protect endangered greenback cutthroat trout around Rocky Mountain National Park, no stocking would be allowed in that zone. Zones around reservoirs on the plains, far from streams with wild trout, relied on stocking for public fishing—they would continue to be stocked. Tony was able to build consensus with this plan, and it was adopted by the DOW.

By listening to his opponents and determining how he could satisfy their needs while also achieving his organization's objectives, Tony continued the process that had led to the defeat of Two Forks Dam—and he helped add a new chapter to the story of the environmental movement's maturation.

WHILE THE WHIRLING DISEASE fight was heating up, Tony convinced TU that getting rid of Windy Gap Dam would be a big win for the organization. In 1998, TU had its first meeting with Northern Water to bring to its attention the declining fishery in the upper Colorado River. And while Barry's whirling disease research at Windy Gap was getting started with funding arranged by Bud, Tony found additional money through TU. Barry's finding that Windy Gap Reservoir was an incubator for whirling disease helped TU make its case that Colorado River rainbows could soon cease to exist—unless serious steps were taken to reverse their decline.

Tony explains he didn't want to back Northern into a corner. "I wasn't interested in assigning blame," he says. His goal was to get everyone to acknowledge the problem and recognize that they had to work together to solve it. He figured if he could get Northern on board with helping solve the problem, then he could find additional funding from other sources.

"The people at Northern were good listeners," Tony says. "They wanted to help solve the problem." Northern agreed to hold more meetings to explore potential solutions. And they were willing to have their engineers study how a bypass channel could reverse the damage done to the river by the dam. Bud and TU developed a draft Memorandum of Understanding with Northern in 2000.

There seemed to be hope for the ailing stretch of the Colorado River below Windy Gap. But then two things happened to stall the bypass channel study.

First, 9/11 and the dotcom bust led to a lot less money being available. Studies of how to heal the damaged river were trimmed from budgets as agencies tightened their purse strings in the austere years that followed the economic downturn.

Second, Barry's whirling disease research on the Colorado River revealed a surprising result: Spore counts below Windy Gap Dam plummeted in 2001. As this trend continued over the next two years, Barry wondered if whirling disease had self-corrected in the reservoir.

Barry explains that the different worm species that can be part of the whirling disease parasite's lifecycle fall into two categories: good and bad. Bad worms host the parasite, allowing TAM spores to spread through lakes and streams and infect trout. Good worms act as biological filters by deactivating spores.

And the good news is this: The bad worms have their reproduction impaired when infected by the parasite—allowing the good worms to outcompete the bad worms and replace them over time. It seems this detail embedded deep in nature's design can sometimes curb total catastrophe caused by whirling disease.

Northern was reluctant to fund a study of a bypass channel if the problem in the reservoir had run its course. Because Barry's research showed that the virulent whirling disease incubator of Windy Gap Reservoir had become more benign, momentum for studying how to heal the river stalled, and a hiatus ensued on the upper Colorado.

Barry turned his attention to the effects of whirling disease on cutthroat trout in Colorado's high-country waters, and Tony focused TU's resources on protecting wild trout throughout the state.

Bud, whose friends and family had been kidding him that his river advocacy work seemed to be his full-time job, focused on running his oil and gas business. While concentrating on making a living during the pause in the conflict, he found time to walk the banks of the river and wade in its waters in search of fish.

Rainbow trout did not reappear. And Bud and others who frequented the river noticed the brown trout population below Windy Gap Dam was also crashing—strange and unsettling given the resilience of the brown trout species. Something was drastically wrong.

When Northern Water proposed the Windy Gap Firming Project, yet another scheme to remove the river's flow, the battle to restore the health of the Colorado headwaters began anew. And Barry, Tony, and Bud resumed their struggle to save the source of this endangered river.

Chapter 11
DEATH BY SILT

When the Windy Gap Project was conceived in 1967, it was said to provide "water for the future." But nearly half a century later, the region that drinks from the Colorado-Big Thompson (C-BT) and Windy Gap Projects is again running dry. One plan to help bail northeastern Colorado out of water bankruptcy would make more reliable (or "firm") the amount of water diverted from Windy Gap.

Northern Water's grip on Windy Gap is weak due to wonky details of infrastructure and water law. Flow diverted from the Colorado River at Windy Gap must be stored in Lake Granby before delivery via the C-BT system to the East Slope. Because Windy Gap water has junior rights, only in wet years can the water be diverted to Lake Granby—posing a challenge for Northern. But what really sets heads spinning is this: When the state is blessed with abundant moisture, Lake Granby fills up, leaving no room for Windy Gap water—creating a catch-22 for Northern.

So, to firm up its grip on Colorado River water slipping through its grasp, Northern Water's Municipal Subdistrict proposed building a reservoir on the East Slope to hold Windy Gap water. In the foothills outside Fort Collins, the Chimney Hollow Reservoir would store water diverted from Windy Gap, firming the supply for the northern Front Range. If the Windy Gap Firming Project comes to fruition, shopping malls on the northeastern plains will not run out of water, subdivision lawns will shimmer greenly in the summer sun, and housing developers will bank more money. The Rawhide Station, a coal-fired power plant near Fort Collins and one of northern Colorado's largest pollution sources, will also use the diverted water. And the source of one of America's most magnificent rivers will be further depleted.

The proposed Windy Gap Firming Project brought long-simmering controversy over the upper Colorado to a boil. The project's permitting process required mitigation plans be assessed by the Colorado Wildlife Commission—a board appointed by the governor and responsible for directing Division of Wildlife (DOW) policy. Before the commission could weigh in on mitigation plans, the DOW needed scientific studies of the upper Colorado River.

Biologists were reporting the Colorado River's brown trout population was following the plight of the river's rainbows. Bud also sounded the alarm. He saw that fish were disappearing, and he noted that many insect species were already gone. He watched clouds of silt stirred up in Windy Gap Reservoir move like choking dust storms through the water below the dam. In spaces between cobbles, the silt collected. In gaps between gravels, it stuck. Cobbles and gravels that had supported life vanished beneath a barren surface as hard and smooth as concrete.

Anyone who spent time on the upper Colorado could see that the natural system was failing. But to create a clear picture of what was happening to the river, research had to be done. Once again, biologist Barry Nehring played a vital role.

BARRY AND HIS CREW WERE tasked by the Colorado DOW with evaluating the river's health. They focused their research on an insect species intensively studied during the 1980s. *Pteronarcys californica*, commonly called the giant stonefly, salmonfly, or willowfly—and sometimes referred to as the pillowbug by locals who remember it—once filled the sky above the upper Colorado, forcing drivers near the river to slow down and use their windshield wipers. Crushed bugs formed a paste so thick on the glass that no amount of wiper fluid could wash it clean. Motorists reached out side windows to squeegee windshields with their hands as they inched down the road through insect blizzards. On the Colorado's banks, exoskeletons shed by giant stoneflies dangled from bushes like Christmas ornaments. Now, the silhouettes of giant stoneflies rarely glide across the sky near Windy Gap, and their husks seldom decorate streamside willows. The crackle of their shed skins beneath boots has been replaced by silent footfalls when anglers walk the shores.

In twenty years the passenger pigeon went from blackening the sky by the billions to zero. The web of life from which a species has been severed is diminished in subtle but important ways. Giant stoneflies disappearing may not seem like an event to be bemoaned—their absence might even seem a boon. Do you want a bunch of bugs covering your windshield while you're trying to drive? And when a giant stonefly lands on your shoulder, this insect seems the size of a small bird or bat; it sends shivers along your spine as its sticky legs ascend your neck with muscular pushing and its inch-long antennae tickle your skin.

But these stoneflies of unsettling size are a crucial food source for fish, and biologists consider the species a key indicator of a stream's health. When giant stoneflies are present in great numbers, fish grow fat and healthy as they gorge on these high-calorie meals; when giant stoneflies disappear, trout turn skinny as they sip smaller insects. It takes an awful lot of pinhead-sized midges to equal the caloric feast of a single giant stonefly.

Trout may be of little interest to someone who lives in a city far from mountains and thinks fishing is as fun as filing taxes. But trout are more important than serving as quarry for anglers—they are creatures finely tuned to the cold waters of streams fed by snowmelt. When trout disappear, something has gone wrong in a river. And most of us in the West drink from rivers that begin in mountains that funnel melted snow to the cities below. Trout and aquatic insects are canaries in the coal mine.

As a river loses its biodiversity, so do the banks surrounding it, and the valleys containing those banks, and the ecosystems that encompass those valleys—and where it will end we do not know. What we do know is that everything is connected: This is the central tenet of ecology. John Muir, after gathering wisdom from the wilderness, wrote, "When we try to pick out anything by itself, we find it hitched to everything else in the Universe."

To see the big picture of the upper Colorado, Barry's crew squinted their eyes in search of insects. They focused on the river reach that begins below Windy Gap Dam and runs downstream to the lower end of Gore Canyon west of Kremmling—more than forty miles of river that once boasted a superb trout fishery. Supporting the trout were what biologists call "aquatic invertebrate fauna": the little creatures trout eat in a stream. This captivating zoo comes alive beneath a magnifying glass—or under the curious gaze of a child who stares at mayflies riding a river's current like miniature sailboats, and who watches a stonefly escape its skin like a magician shrugging off a straightjacket.

Barry explains that when a giant stonefly metamorphoses from larva to adult, it crawls from the water to latch on to a solid surface with its clawlike "toes." Once locked in place, the larva arches its back, splitting its exoskeleton on the dorsal surface of the thorax—like our back, notes Barry. The giant stonefly flexes its back, popping out its wings, which unfurl and dry. Then the adult bug crawls out of its skin.

Barry's manic enthusiasm for the creatures that live in rivers, coupled with his remarkable physical stamina, make him seem more like a hyper boy than a retired biologist in his eighth decade. Rachael Carson, best known for her book *Silent Spring* that ignited the environmental movement in America, also authored *The Sense of Wonder*, in which she wrote, "A child's world is fresh and new and beautiful, full of wonder and excitement. It is our misfortune that for most of us that clear-eyed vision, that true instinct for what is beautiful and awe-inspiring, is dimmed and even lost before we reach adulthood."

To the benefit of everyone who cares about wild trout and healthy rivers, Barry and his colleagues never lost their childlike wonder at the many forms of life that inhabit the waters of the West.

Barry heard Bud speak of giant stoneflies disappearing at Chimney Rock Ranch downstream of Windy Gap, but he was skeptical. Most fly fishers are not professional entomologists. Over the years, Barry had heard many anglers confuse insect species—understandable given the astounding diversity of aquatic insect life. What we call "stoneflies" comprise some 3,500 separate species.

When the proposed Windy Gap Firming Project led to Barry being asked to repeat a study of the giant stonefly species in the upper Colorado that he'd done more than two decades earlier, he decided to find out if Bud's story was true. He broadened his sampling survey to include sites on Chimney Rock Ranch and in the Fraser River upstream of Windy Gap.

Barry expected to find giant stoneflies in the Colorado below Windy Gap in roughly the same numbers as at his sampling sites farther downstream. To his surprise, on the river reach below Windy Gap, the giant stonefly had nearly vanished. Bud had been right.

Barry's crew found giant stoneflies in Willow Creek and the Fraser River above Windy Gap Reservoir—indicating Windy Gap Dam had caused their disappearance downstream. But this was just an educated guess. A crucial piece of information was missing.

Science is a detective story. Detectives rely on logic to gather and piece together clues, but chance often plays a pivotal role.

Barry's stonefly research wouldn't have amounted to much if not for a lucky break.

To make a solid case that Windy Gap Dam had decimated insect populations in the river downstream, Barry and his team needed a baseline study of buglife prior to the dam's construction. No such study had been published. Fly fishers like Bud spoke of giant stoneflies filling the sky below Windy Gap in years past. But to do sound science, Barry needed data, not anecdotes.

While thumbing through a binder one day, Barry noticed a handwritten letter with unpublished data about giant stoneflies below Windy Gap in 1980 and 1981, before the dam's construction. The letter had been sent to Barry in 1982; he'd filed it away twenty-seven years ago. Barry tracked down the phone number of the scientist who'd compiled the data, Dr. Robert Erickson, and gave him a call.

The day Barry spoke with him, Dr. Erickson was cleaning out his garage—he was just about to throw his unpublished report in a trash bin. He sent Barry the report. "Getting hold of that unpublished study was like coming across the Dead Sea Scrolls," says Barry.

Dr. Erickson had led a team of scientists to gather detailed data on lifeforms back when the river still flowed freely past Windy Gap. The study documented the relative abundance and distribution of dozens of aquatic insect species, including *Pteronarcys californica*, the giant stonefly. This study was the missing piece Barry had sought. But the serendipity of this discovery was not enough to solve the puzzle of the river's decline.

Dr. Erickson's research had been funded by Northern as part of the environmental impact statement (EIS) process for permitting Windy Gap Dam, but had never been made public. Because Northern's project was approved before the study concluded, there had been no need for the agency to release the report. When Bud got wind of what was happening, he had his attorney file a Freedom of Information Act to prompt Northern to release the study—and Dr. Erickson's research became publicly available.

Barry now had baseline bug data to which he could compare a current census. He scrupulously repeated the insect sampling methods used in the 1980–1981 study. Barry even hired Dr. Erickson for a day to make sure he was sampling insects in the same riffles where Dr. Erickson had conducted his study.

The results were telling. The 1980–1981 "before" study and the 2010–2011 "after" study provided evidence for an airtight case that Windy Gap Dam was killing the Colorado.

BARRY EXPLAINS THAT THE presence of many giant stoneflies during a stream census tells scientists a river is functioning as it's supposed to; declining numbers of this species indicate a river's failing health. The complete absence of giant stoneflies in a reach where they were once abundant indicates an ecosystem is, as medical professionals say of patients out of earshot, circling the drain.

The giant stonefly fits the bill for an ideal indicator species. Large and easy to identify, it occurs in many rivers throughout the American West, though it requires very specific conditions. Sluggish flows and silty streambeds don't support the giant stonefly; its larvae thrive in fast-flowing, highly oxygenated mountain rivers with steep gradients and cobble streambeds. Finding these conditions on a river but no giant stoneflies is like seeing an oak tree covered in acorns but no sign of squirrels: Something is amiss.

A study of the upper Colorado River Barry ran for eight years beginning in 1982 predicted silt accumulation in the streambed below Windy Gap Dam would threaten the stonefly's survival. Barry concluded in his report, "The Windy Gap Project is just one more incremental assault on riverine aquatic habitats in the upper Colorado River basin that have been going on in Colorado for more than half a century. And it certainly will not be the last."

When Barry and his DOW crew[16] conducted the 2010–2011 study, giant stonefly larvae were almost totally absent in riffles where thirty years previous they had been abundant. The giant stoneflies Bud had once seen triggering trout feeding frenzies when swarms of the insects crashed into the water had been so devastated by changes to the river, hardly a single one remained.

The 2010–2011 study revealed further disturbing developments on the river. Other important aquatic insects, including two species of large mayflies present in the 1980–1981 study, had also gone missing below the dam. These three species were, however, still found on the Fraser River upstream of Windy Gap Reservoir—supporting Barry's hypothesis that Windy Gap Dam was to blame for the demise of these creatures. Barry wrote in the report this call to action: "Storage, impoundment and diversion of the waters of almost all creeks and major streams in Grand County have created an untenable situation that will soon be impossible to correct unless a massive investment and commitment of time, manpower and capital is forthcoming on a number of fronts."

Barry's study documented the disappearance of 38 percent of aquatic insect species over two decades, and he concluded that chronic sedimentation caused by the dam was the main culprit. Silt was the smoking gun.

---

16   Barry emphasizes that the research was a team effort. He says, "The work I've done for the state almost always involved a crew of three to six people or more. I have been amazingly blessed to surround myself in my work with incredibly gifted, smart, talented, hard-working people who shared my passion for developing a thorough understanding of how aquatic organisms interact with their environment and are affected by it."

The 2011 report has three authors: Barry is listed as principal investigator, with Brian Heinold and Justin Pomeranz as coauthors. Barry says, "Brian and Justin did all the heavy lifting on the bug study component of that report—including all sorting, identification, synthesis, analysis, and writing on virtually all the bug data, except for the stuff that deals directly with *Pteronarcys californica,* the giant stonefly."

~~~~~~~~~

FROM THE DAY A RIVER is dammed, the reservoir it forms is
doomed. When a river's path is blocked by a dam, the river
deposits the load of silt it carries on the floor of a reservoir.
How soon a reservoir turns from a water container into a tub
of mud depends on the reservoir's size and how much sediment the river transports. In the case of Windy Gap—or Muddy Gap, or Windy Crap, as some people call it—the reservoir
is small, the river's sediment load large. Windy Gap's volume
decreased by half as silt settled to the shallow reservoir's floor.

Aside from compromising the storage capacity of reservoirs,
silt carried by the Colorado can damage aquatic life when the
river's hydrology is altered by water projects. After Windy Gap
Dam was built, the river below it began to die.

Across the world, the manipulation of rivers by damming
them, dewatering them, and changing their natural flow patterns has decimated aquatic organisms and driven species to
extinction. In rivers downstream of dams, a decrease in biodiversity has been documented around the planet. Nonetheless,
Northern has argued that just as much biomass exists now
below Windy Gap Dam as before it was built, raising the ire
of Barry and other biologists. They point out that even if total biomass is maintained downstream of a dam, diversity is
diminished—and when native insect species disappear, repercussions to the river's health can be devastating.

Aquatic insects are the currency that runs a river's natural economy. Insects consume organic matter in a river, converting raw material into energy that passes through the food
chain as the insects are fed upon by fish and frogs, by birds and
bats. Insects link a river's flow to its banks and skies, joining
water, land, and air in one interconnected system of borderless
countries through which these tiny beings travel, circulating
in a globalized economy of energy transfer.

Among aquatic insects there is remarkable diversity. At the
risk of oversimplifying the complex science of entomology, a

few generalities can be made. The lifecycle of many aquatic insect species in temperate zones, including the state of Colorado, lasts one year. Most of this time is spent in the water in the larval stage. Insects feed and grow in the river's flow and then rise to the surface, emerging from the water for a few hours to a few weeks. In their adult stage, the last of their brief lives is given to reproduction.

Aquatic insects spread through a river mainly by drifting downstream. As larvae, they are carried by a river's current, both when their grip on the stream bottom is disturbed, and when they detach themselves to float in search of better habitat. Downstream drift increases at night and is highest during dark phases of the moon. Tiny travelers beneath starlit skies, aquatic insects release their hold on rocks and tumble in the river's flow. When their journey is blocked by a dam, they cannot colonize reaches downstream, altering the river's ecosystem.

This broken connectivity is one factor in the loss of insect biodiversity in a dammed river. Sediment is another.

With the health of a river like the Colorado, not all sediment is created equal. Similar to good cholesterol and bad cholesterol, sediment is essential for the healthy functioning of a stream but also can cause its death, depending on the size of the grains and how they move through the system. Coarse sediment collects behind a dam: Cobbles and gravels sink in the reservoir like nuggets in a gold miner's pan—instead of spreading downriver to create good streambed habitat for insects and fish. Fine sediment, or silt, remains suspended in water that passes the dam. Downstream of Windy Gap, silt settles, filling in spaces between stones in the streambed. Barry calls these spaces "condos" for aquatic insect larvae, especially mayflies and stoneflies.

In a healthy, undammed river that swells with spring snowmelt, flushing flows scour the streambed clean. High waters blast away silt, leaving scrubbed cobbles and gravels in the wake of each regenerating flood. Without these "channel maintenance flows," silt accumulates on the streambed, cementing stones together as it cakes in hardened layers.

Bud puts it bluntly: "It's just like flushing a toilet. How long do you leave crap in the toilet before you flush it?"

Most of the water diverted from the upper Colorado River Basin is taken from runoff, making spring flows too weak to flush the river clean. Windy Gap creates an additional problem. Because the reservoir is so shallow, fluctuating water levels disturb sediment on the reservoir's muddy bottom, and stirred-up silt spills from the dam. When the reservoir is drawn down in summer and refilled, sediment pours into the shrunken river below. Bud once watched water the color of creamed coffee discharge from the dam and spread in dark plumes downstream. The river's summer flow was too feeble to transport this sediment, which clogged the streambed and hardened over time—similar to plaque building up in the arteries of a person suffering from coronary heart disease.

The longer the river goes without a good strong flush, the further its health declines. Starved of gravel and engulfed by silt, the river slips toward a sterile state in which life cannot thrive. To the insects that live among pebbles and stones of a streambed smothered by fine sediment, it is as though a volcanic eruption blanketed our world with several feet of ash. The Colorado below Windy Gap is a silt apocalypse—a Pompeii for the little creatures of the riverbed.

And trout, besides being deprived of crucial food sources by silt-smothered streambeds, cannot spawn: To successfully reproduce, they need clean gravel. The more water the Front Range siphons from the Colorado's spring flows, weakening their purgative power, the more the river suffers. Weeds make the problem even worse.

After Windy Gap Dam was built, plants filled the reservoir's sun-warmed waters with green and swaying strands. About three feet in average depth and choked with weeds, Windy Gap Reservoir seemed more like a swamp in the Everglades than an alpine reservoir in the Rockies. Bits of plants broke off and drifted downstream of the dam, where they rooted in the sluggish river. With no flushing flows to dislodge them, the plants spread in thick mats, blocking the river's current. These weed mats filtered silt from the water, further filling the gaps between cobbles and gravels.

The dammed river downstream of Windy Gap also became perfect for breeding a species of algae known as didymo, or "rock snot." This slimy green goop makes river rocks slippery as snot and sticks to the legs of people wading through cloudy waters. Algae blooms rob oxygen from rivers—without sufficient oxygen, insects and trout cannot survive.

From a cold and clear-flowing stream full of trout that gorged on giant stoneflies, the Colorado River below Windy Gap in drought years became a warm and weed-choked trough of muck. Not only did Northern fail with its Windy Gap Project to provide a firm yield of water to its customers on the Front Range, it also created a muddy reservoir that ruined the river. "They get an F for the project," Bud says as he shows me photos he took of Northern using construction equipment to dredge the sludge from Windy Gap Reservoir to protect its pump plant. This dredging stirred up silt that choked the stretch of river Bud watches over.

Aldo Leopold wrote, "One of the penalties of an ecological education is that one lives alone in a world of wounds." Damage to the Colorado River's streambed, smothered with silt and deprived of flushing flows, might not be obvious to the untrained eye. But to Bud, who for many years waded in the waters and turned over rocks to look for insects, the transformation caused by Windy Gap Dam and too much flow being diverted from the Colorado headwaters was like going to a favorite park and seeing the grass paved over with a parking lot.

And the condition of the upper Colorado River could get even uglier. According to a study by Resource Engineering, Inc., if projects proposed by Northern and Denver Water proceed as planned, May streamflows below Windy Gap would be reduced by an additional 35 percent, further weakening the river's power to purge itself.

To Bud, it became obvious a long time ago that intervention is needed to flush away the silt that entombs the upper Colorado. He realized that reversing the river's decline requires decisive action.

Barry was reaching the same conclusion as his research on the river progressed.

A RIVER PERIODICALLY NEEDS to be "exercised," Barry explains to me as we hike down the Gunnison Gorge near his home. We head toward a green ribbon of water winding through the canyon, stepping past bright slabs of sandstone, dropping through dark bands of volcanic rock made slippery with skittering balls of hail. Wizened bushes lean away from the wind that pummels ridgelines, scattering the smell of gin from crushed juniper berries. A few trees hunker in sheltered nooks. Barry and I grab their bonsai branches for balance as we lower ourselves past boulders cracked in ancient cataclysms.

When we reach the canyon floor, we stand squinting against a hailstorm at the river's edge, studying glassy pools and foaming rapids. Barry explains that flushing flows in a river roll the rocks around, breaking apart silt that concretes the streambed—loosening everything up until a clean cobble bottom replaces hardened mud. "A spring flood is like a rototiller in a garden," he says. I stoop to pick up river stones smooth as eggs. Beneath the water's surface, I scratch at a sheet of silt soon to be scoured by spring floodwaters.

Unlike the upper Colorado, the Gunnison in Barry's backyard is blessed with spring floods. These floodwaters remain in the river due to a legal tangle of water rights complicated enough to allow attorneys to pay off their student loans and put their kids through college. Floods will flush the Gunnison of silt, allowing its refreshed bed to provide food and shelter for the insects that fatten trout. The Colorado downstream of Windy Gap, conversely, will be teased with a trickle of water— unless the river management regime is changed to provide big spring flows that blast the riverbed clean.

As Barry and I kick over cobbles to look for bugs, he explains that a river's natural flow pattern signals to insects when biological changes should take place—when eggs should hatch, for example. Many aquatic insect species evolved with their lifecycles tuned to subtle variations in stream temperature. When a river is deprived of water in hot summer months, its reduced volume quickly warms. In a shallow basin like Windy Gap Reservoir, the sun's radiation heats water that then flows downstream: The solar collector of the reservoir creates thermal pollution.

Changes in temperature cues can wreak havoc with insects that evolved to survive natural selection pressures in a watershed—pressures that didn't include thirsty suburbs sucking the Colorado down to a hot trickle. Water temperatures that rise too high can also be deadly for trout, which evolved in the cold waters of snow-fed streams. As stream temperatures rise, the amount of dissolved oxygen in the water decreases; without enough oxygen, trout die. They are creatures finely tuned to their environment, as are we. When a river's flow pattern is altered, the world of aquatic insects and trout is as radically changed as ours would be if the sun dimmed, the seasons stopped turning, and oxygen disappeared from the atmosphere.

Scientists call the seasonal rise and fall of a river's flow its "hydrograph." Graphs of stream levels in healthy watersheds show jagged lines that spike and dip in a yearly cycle. This natural flux is crucial to the healthy functioning of aquatic and riparian ecosystems. When a river is starved of peak spring runoff, floodwaters don't spill over the banks. And species from songbirds to moose suffer as their habitat is degraded and the river's hydrograph "flatlines" with a deadly lack of variation.

Alison Holloran, a research scientist and executive director of Audubon Rockies, explains that riparian habitat is an "ecotone": a transitional area where two communities meet, where water laps against land. In the science of ecology, edges are where the action is. Riparian habitat represents a mere 1 percent of the land in western states, yet 90 percent of all the bird species in the Colorado Plateau region rely on riparian areas for food, water, cover, and migration corridors. Riparian habitat also nurtures many reptiles, amphibians, and mammals. When we deprive a river of spring floods, we damage the structure of riparian zones and impact their plant life, harming birds and other creatures in a cascade of consequences that spreads beyond riverbanks. According to Alison, "Healthy habitat equals healthy ecosystems equals healthy humans."

When Alison takes off her scientist's hat, she talks about the poetry and power of the sandhill crane. Each time she hears its trilling call she is transported back to a time in her life when she heard the calls of cranes on the Wyoming plains and everything was going well for her professionally and personally. The sound of the sandhill crane elicits for her those rare moments of harmony when everything in the world seems aligned as it should be. In wet pockets on the prairie, this elegant bird dances with gangly grace and then takes to the sky, the rattling bugle of its call finding its way to Alison's ears. Her argument about keeping water in rivers to support the many forms of life that rely on riparian corridors is scientifically sound. But her hunger to hear the calls of cranes is also noteworthy. Wild places enrich our lives in ways that cannot be measured with the tools of science, and wild creatures uplift us in ways impossible to quantify.

When proponents of diverting the Colorado River headwaters to Front Range faucets and lawns insist their efforts are being impeded by a handful of fishermen who want to catch trout, they are disregarding the vast webs of life that emanate from the river's flow. And they are ignoring the incalculable solace and joy these living systems provide. Putting a price on the ancient call of sandhill cranes filling our ears is not possible. Nor is it possible to know the consequences to the human heart if species that evolved through millennia of the earth's story disappear, and the mountains and prairies fall silent. But we can say with certainty that if we turn rivers to trickles, our world will be impoverished in ways we cannot predict.

When water providers reassure the public they are "only" skimming excess from a river overflowing with spring runoff, causing no harm to the river's health, these professionals are either ignorant of the science of watershed ecology, or they are being disingenuous. I have never met a water manager who wasn't very sharp. Theirs is a complicated and demanding job, and only the best rise to the ranks where they must explain to the public why they need to further drain what's left of our rivers.

As knowledge of water law is essential to making sense of why rivers are being killed to give life to luxurious lawns and subsidized sugar beets, so is scientific literacy crucial when analyzing the questionable claims of water providers, some of whom aim to dupe an uninformed public. Depriving the Colorado of spring flushing flows has damaged the river—as demonstrated by the disappearance of the giant stonefly documented in scientific studies.

To further prove the problems facing the upper Colorado River were enormous, Barry focused on a little fish.

Chapter 12

RESTORING THE RIVER

Though not classically attractive like the rainbow trout, the mottled sculpin is beautiful nonetheless. The beauty of this native fish species lies in its expression of nature's efficient design. If the rainbow trout is the Sistine Chapel of fish, then the mottled sculpin is a work of contemporary art that challenges us to appreciate its functional aesthetics.

The mottled sculpin belongs to a family of fish called Cottidae that occurs all over the planet in rivers, creeks, and seas. At first glance, the mottled sculpin looks like a small version of its ocean relatives, but the species is perfectly adapted to the environment in which it evolved: water born from melted snow that flows over sandy riffles and beds of gravel. The fish's mismatch of stripes, spots, and speckles—what biologists call "cryptic coloration"—provides perfect camouflage against the pebbles of a mountain stream, where it stays in place for long periods and then darts in a strange motion that seems more like hopping than swimming.

When you cup the moist weight of a mottled sculpin in your hand, the feel of its slippery skin without scales against your palm, combined with the look of its oddly flattened body and

oversized head, make it seem less like a fish than a tadpole—
or a transitional being from dim prehistory when creatures in
swamps first made their way awkwardly onto land.

The anatomy of the sculpin, an archaic species of fish with
no air bladder, makes it a weak swimmer but a strong clinger.
Pelvic fins that fan roundly at the edges stretch across a scaf-
folding of soft, flexible spines that allow the little fish to hold
itself steady on the bed of swift-flowing streams. With its agile
fins, the sculpin anchors its body between cobbles as it feeds
on insects that creep among stones or drift along the stream
bottom.

The mottled sculpin feeds mainly on aquatic insect larvae
that dwell in riverbeds; in turn, the sculpin is preyed upon by
trout. Remove aquatic insects from a river and sculpin disappear. Remove sculpin and trout struggle. Take away trout
and the osprey that feed on them vanish. An ecosystem is a
three-dimensional puzzle that challenges us to remove pieces
without toppling the structure, which is held together by the
precarious tension of each piece resting against neighboring
pieces. Pull out one creature, whether the giant stonefly or the
mottled sculpin, and the whole thing starts to fall.

The mottled sculpin's world consists of gaps between gravels and cobbles. In these small spaces the little fish ambushes
insects and hides from predatory trout. When these spaces fill
with silt not flushed away by a flood, the streambed smothers
beneath a crust of mud—and the sculpin disappears from its
native waters.

Like the giant stonefly, the sculpin is an excellent indicator
species. Locating sculpin in their limited home range, no larger than the carpet in Barry's living room, he tells me, is much
easier than chasing mobile trout. Because sculpin are extreme-
ly sensitive to metals in streams—even miniscule amounts of
zinc can kill them—the species is an effective bioindicator of
water contaminants. And the spawning nests and incubating
eggs of sculpin are highly sensitive to changes in streamflow.

A large sculpin population in a stream indicates to
biologists the water is clean and the ecosystem is healthy. In

short, if sculpin are present, the river is in good shape. If sculpin are gone, something is wrong.

Barry says, "I witnessed this little fish disappear from the Colorado River below Windy Gap over a period of twenty years."

The 2011 Division of Wildlife (DOW) study Barry led documented the mottled sculpin's distribution throughout the upper Colorado River watershed. The sculpin was the most abundant fish species in every Colorado River tributary, including the Fraser River—right up to the Fraser's confluence with the Colorado above Windy Gap Reservoir. Below Windy Gap Dam, all the way downstream to where the depleted Colorado is bolstered by the Blue River, the sculpin had disappeared. This sentinel species that evolved over millions of years to be seamlessly fitted to its environment could not survive the silt that clogged its streambed habitat. Nor could the mottled sculpin adapt to the artificial fluctuations in the river's flow caused by pumping and diversions.[17]

---

17   Barry elaborates in damning detail: "The sculpin is the most abundant fish species in every tributary of the Colorado River above all the dams in Grand County—Windy Gap, Granby, Shadow Mountain, Willow Creek, and Williams Fork. This is true even of the Williams Fork River upstream of Williams Fork Reservoir right to the high-water line of the reservoir—but the sculpin is gone from the two- to three-mile reach of the Williams Fork River downstream of the reservoir to the confluence with the Colorado River, except for a few fish in the Williams Fork River immediately below the dam that came through the outlet tubes.

"With this prima facie evidence, it's pretty hard to refute that the only reason for the disappearance of the native sculpin is the long-term effects of manmade manipulations of the natural hydrograph and withdrawals of water on this fish. You can't lay it on climate change, alteration of the thermal regime, water pollution of any kind, or anything else. This is especially true when you know that below the Blue River augmentations for servicing downstream water rights to the Shoshone Hydro Plant and Grand Valley, the sculpin again is the most abundant species through Gore Canyon all the way to Radium and beyond."

Barry explains that creatures in aquatic and riparian environments not only need water to survive—they require stable seasonal flow patterns to successfully reproduce.

The mottled sculpin has a complicated sex life. The male builds a nest tucked under cobbles; then he courts the female to join him. Once inside this nest, which Barry calls "a sort of nuptial grotto," the female swims upside down and attaches her sticky eggs to the ceiling. Her male suitor fertilizes the eggs and then guards them. After hatching, the baby sculpin drop to the floor of the nest, and papa sculpin protects them against predators until they swim away in search of food among the nooks and crannies of cobbles. The whole process is controlled by water temperature and usually lasts for six to eight weeks during the late spring and early summer months—when large amounts of water are removed from the river at Windy Gap.

When river diversions during the spring reproductive period dry up the nest, egg incubation and embryonic development fail. And the sculpin vanishes from the rivers and streams that gave the species life through the millennia, as mountains rose and were whittled down by weather, and as water carved channels filled with lifeforms reliant on seasonal streamflow. In a geologic eye-blink, the sculpin is gone. Like the bison and the beaver, the grizzly and the wolf. And perhaps like us if we continue to destroy the rivers upon which our survival depends.

Noted by Bud and Tony through their on-stream observations, and confirmed by scientists in exhaustive studies, the health of the Colorado below Windy Gap was, in layman's terms, going downhill on a greased slide.

BARRY CONCLUDED THAT DAMS and diversions in Grand County had pushed the ecology of the upper Colorado River watershed dangerously close to collapse. The damage had begun back when the Grand Ditch was carved to divert water from Grand County across the Continental Divide, and when Denver Water's Moffat Tunnel diminished the flow of almost every stream tributary to the Fraser and Williams Fork Rivers.[18] The five major dams that blocked the rivers and streams in the upper Colorado watershed altered the region's hydrology. The cumulative effects of these projects strained the river's health, and Windy Gap Dam delivered the coup de grâce, killing off the giant stonefly, the mottled sculpin, and a whole host of other creatures in the upper Colorado. Further diversions from Northern's Windy Gap Firming Project and a proposed project of Denver Water threatened to make the situation even worse—perhaps giving the river the final shove that could turn it from dying to dead.

In an early draft of Barry's report, he compared Grand County to the Owens Valley, plundered of water by duplicitous powerbrokers in Los Angeles. Barry noted that unless "serious attempts" are made to address the deterioration caused by the dams and diversions in Grand County, "it will not be long before the Colorado River in Grand County is as tapped out as the Owens River in California."

Whether Los Angeles snatching a river from the Owens Valley is analogous to Front Range water providers drying up Grand County is a matter of spirited debate. Regardless, parallels between the emptied Owens River and the tapped-out Colorado are apt—both in their destruction and in their potential revival.

When water is once again allowed to flow through depleted channels, rivers soon experience a resurgence of life, as with the Colorado River Delta mentioned in the prologue. In the case of the Owens River, after Los Angeles was forced by court order to reverse some of the damage it had caused, the city embarked on what the *Los Angeles Times* called "the largest river restoration effort ever attempted in the West." In 2007, water was redirected back into the riverbed. And from this wetness life returned. After almost half a century of absence, willows again rose from the riverbanks, providing shelter from the sun, and stands of cattails grew busy with crayfish, frogs, and

---

18   Denver Water's depletion of the Fraser River is explained in later chapters.

bass as herons patrolled the shores.

Though the Colorado below Windy Gap has been harmed, Barry remains optimistic it can heal. He is confident mottled sculpin, giant stoneflies, and wild trout can once again thrive in its waters. This hopeful scenario is not without precedent.

In Grand County's Pioneer Village Museum, a display reads, "In 1854 Sir St. George Gore wandered through the area, killing animals by the hundreds, leaving little but his name." Tim Nicklas of the Grand County Historical Association says old-timers speak of hunting seventy years ago being poor in the Colorado River headwaters region. Meat hunters decimated Colorado's elk to feed mining camps. But through careful stewardship, herds increased from a handful of animals that survived this carnage, and Colorado now boasts more elk than any other state. When you drive Highway 40 in winter near Windy Gap, you can see steam rising from the backs of antlered beasts that migrate down from the high-country into hay meadows.

Cutthroat trout once overflowed from the creels of Coloradans. Along with commercial fishermen that fed hungry boomtowns, recreational anglers stripped this bounty from the state's rivers in a few decades. With bright yellow fins and a lemon hue along its sides, the yellowfin subspecies of cutthroat trout grew up to ten pounds. Now extinct, the yellowfin cutthroat exists solely as preserved specimens in the Smithsonian Museum. But a small population of greenback cutthroat trout, Colorado's state fish, survived in a single stream. Now aggressively protected, the greenback cutthroat is being returned to its native waters. Where there is a will to save rivers and wild creatures, there is always hope, Barry insists.

Restoring the river will take as much effort as building Windy Gap Dam demanded. According to Barry and other biologists who have commented on upper Colorado mitigation and enhancement[19] plans, for the river to be re-

19   In river management jargon, "mitigation" refers to reducing the impacts of a water project; "enhancement" means improving the quality of a river that has already been impacted by a project. Put more simply, "mitigation" protects, "enhancement" restores.

stored to health, a free-flowing bypass must be created. A channel around Windy Gap Reservoir would reestablish the river's connectivity, allowing critical biological functions to resume. Mottled sculpin and giant stoneflies in the Fraser River could recolonize the Colorado River below Windy Gap, along with green drake mayflies and other aquatic insects missing below the dam. And trout would be able to swim upstream and downstream in search of food and spawning sites.

A bypass channel would transport cool, oxygenated river water instead of the warm, weedy flow released from the reservoir, keeping the river's temperature safe for trout below Windy Gap and reducing the buildup of rooted weed mats. And nutrients, instead of being trapped behind the dam to change the Colorado's chemistry and biology, would move freely in the reconnected system, restoring the balance that evolved over many hundreds of centuries.

Perhaps most important, a bypass channel would solve Windy Gap's sedimentation problem. The channel would take the silt-clogged impoundment off-stream; it would also allow gravels and cobbles that fish need for spawning to spread downriver.

Barry points out that although the prevalence of whirling disease has been reduced, a change in tubifex worm populations, or some other shift in the stream's ecology, could allow the parasite to once again thrive in the reservoir. A bypass channel would reduce the exposure of the Colorado River below the dam to deadly spores.

Along with the creation of a bypass, enough water must be allowed to flow downstream each spring to break up the silt that armors the riverbed, leaving clean cobble beds and gravel bars. Barry and other biologists who have weighed in on restoring the upper Colorado watershed point out that these flushing flows, which mimic the river's natural cycle, are as important as building a bypass.

Barry explains that narrowing the Colorado's channel below the dam is also necessary. A river made skinny by diversions should have its channel resized—like a person who loses weight wearing slimmer pants. Robust rivers that haven't been starved of their flow by diversions form deep pools and swift currents; squeezing an emaciated river between narrowed banks recreates these natural processes.

Barry is confident that with these interventions—building a bypass, providing spring flushing flows, and narrowing the river channel—the Colorado below Windy Gap Dam can be restored to health. A dead streambed of mud will disappear. Clean gravel will glint in the sun, and insects will tumble through refreshed currents to be snatched by mottled sculpin on the river's cobbled floor. Spring floods will blast away the algae and weeds that choke the Colorado's flow, and the hot reservoir will stop spilling into the river. In waters cold and clear, trout will fatten on sculpin and stoneflies, these healthy fish will feast from dozens of other insects that return after decades of absence, and eagles will watch from the trees.

Had a bypass channel around the reservoir been included in the original plans for Windy Gap when the project was being scrutinized in the environmental impact statement (EIS) process, much of the damage to the river below the dam most likely would not have occurred.[20] A free-flowing channel would have kept the river connected and prevented the muddy reservoir from dumping silt, weeds, and whirling disease spores downstream.

But that is water under the bridge, or under the dam, as it were. The task now is to build the bypass—to reconnect the river in a channel that allows water and life to flow past the deadly reservoir.

Bud and Tony almost made that happen fifteen years ago, when they began their battle to restore the river. But, as mentioned, the economic chaos caused by the tech bubble bursting

---

20   A bypass was created as part of the Windy Gap Project, but is much different from what proponents of healing the river are proposing. The existing bypass was designed to deliver a relatively small amount of water from the reservoir downstream to fulfill a water right on property below Windy Gap Dam. The proposed bypass referred to in this book would reconnect the river's natural flow in a channel that completely bypasses Windy Gap Reservoir—effectively doing away with Windy Gap Dam as a factor in the river's demise.

and 9/11, along with Barry's surprise finding that whirling disease spores in the reservoir at Windy Gap were diminishing, had interrupted plans to study and build a bypass.

Barry's latest research, prompted by the proposed Windy Gap Firming Project, revealed that sculpin and giant stoneflies had vanished, providing evidence that the river was in serious decline and a bypass was desperately needed. But there was a catch. There always is.

As the public comment period for the Windy Gap Firming Project EIS was coming to a close in 2011, and as the deadline for the Colorado Wildlife Commission to approve mitigation proposed by Northern Water and Denver Water neared, the Colorado Department of Natural Resources refused to release Barry's study. His research was buried by bureaucracy and hidden by parties that didn't want his science revealed to the public.

But that wasn't the end of the story.

Tony explains there was one wildcard that proponents of the Windy Gap Firming Project never counted on—one surprise that "scared the hell out of them," according to Tony. That was Bud Isaacs. "His determination is only exceeded by his passion for the resource," says Tony. "If I'm going into a fight, I want him on my side."

Tony went to work for the Upper Colorado River Alliance (UCRA), a nonprofit Bud started with his friend George Beardsley, a developer known for his commitment to environmentally sound projects. George's passion for the Colorado River rivalled Bud's. By serving on the Denver Water Board, George saw how the organization operated from the inside. And as a member of Chimney Rock Ranch and as a ranch owner in the Blue River Valley, he witnessed firsthand the decline of the upper Colorado River watershed. Bud and George decided the best way to fight back against the diversions of Northern and Denver Water that were killing the river was to form a coalition. Together they launched UCRA with an ambitious agenda to represent landowners on the upper Colorado River and to restore degraded river reaches, protect the river's health, and enhance fishing and recreational opportunities. The organization's goals also included conducting scientific research, educating the public about the upper Colorado River, advocating for private property rights and water rights, working to create coordination and cooperation among groups involved with the river, and participating in permitting processes.

When George's battle turned from fighting for the river to fighting against cancer, which ended his life in 2011, Bud and Tony took the helm of UCRA and steered the organization into a battle to get a bypass built around Windy Gap Reservoir. By making the case that getting rid of the dysfunctional dam at Windy Gap would be as momentous as the cancellation of Denver's Two Forks Dam, Tony enlisted Colorado Trout Unlimited and national Trout Unlimited in the cause.

The goal of Tony and Bud was clear: build a bypass to restore the river. The science was on their side, and momentum was moving in their direction. But powerful interests pushed back.

Chapter 13
SUBTERFUGE

The vast majority of money in Colorado resides on the Front Range, and that's where water is forced to flow. Through colossal diversion projects that penetrate mountain ranges, rivers are rerouted to cities, wetting the lawns of constituencies that elect the state's most powerful politicians. Sculpin and stoneflies don't vote. And all the votes of Grand County, with its population of some 15,000 souls, pale in importance to the ballots cast by the residents of a single Front Range suburb.

It is not surprising that the Colorado Department of Natural Resources (DNR) wasn't quick to release Barry's report. The Division of Wildlife (DOW)[21] is charged with protecting the state's wild creatures, but is part of the DNR, which has this mission: "to develop, preserve, and enhance the state's natural resources for the benefit and enjoyment of current and future citizens and visitors." Combining "development" and "preservation" in one department is a recipe for confusion, if not conflict.

---

21   In 2011 the Division of Wildlife was combined with Colorado State Parks to form Colorado Parks and Wildlife.

Preserving the Colorado River headwaters makes good sense for sculpin and stoneflies and trout, for moose and birds and bears—and for Coloradans who value healthy watersheds. Developing the state's water resources for the benefit of the Front Range, where most of the votes are, makes good sense for people with political ambition.

⁓⁓⁓⁓⁓

As director of the DNR, Mike King oversaw the Colorado Wildlife Commission's review of the mitigation plans submitted by Northern and Denver Water. He also led the department responsible for releasing Barry's report.

The DNR announced it would release the report in June 2011. But June came and went and the report was not made public. A couple more months passed—nothing but crickets.

Barry's frustration rose. The public comment period for the Windy Gap Firming Project environmental impact statement (EIS) was coming to a close, yet DNR still hadn't released the report.

How could the public provide informed input on the project when it didn't have access to this study? How could the Colorado Wildlife Commission, the entity entrusted with protecting, preserving, and enhancing the state's wildlife and wildlife habitat, make a good decision about mitigation plans for the river without referring to the best science?

Wildlife commissioners explained to Bud and Tony they had been instructed to vote on the mitigation plans based solely on how the plans would address impacts of Northern's Windy Gap Firming Project and a proposed project by Denver Water. Even though everyone knew the river was already severely degraded, the commissioners were told damage done in the past could not be discussed or considered. Tony says, "This was like a jury being told by a court to ignore a crucial piece of evidence when making its decision." Tony's favorite author is Franz Kafka, the creator of nightmarishly illogical tales like *The Trial*—which seems fitting because the bizarre twists of

Tony and Bud's battle sometimes approach the surreal quality of a Kafka novel.

When Bud got word of a wildlife commission meeting in Salida, he contacted DNR to find out if the mitigation plan for the upper Colorado River would be discussed. "Mike King's office told me that the topic wasn't an important part of the agenda," says Bud. "But I smelled a skunk." He went to the meeting anyway. The main topic of discussion was the upper Colorado mitigation plan.

After that incident, Bud made a point of attending all the wildlife commission meetings on both sides of the Continental Divide, traveling as far as Grand Junction to make the case for restoring the upper Colorado River. "The state's namesake river is dying," Bud said in a public comment session at one of the meetings. "We're asking the governor, state wildlife commissioners, and the Department of Natural Resources to uphold their responsibility to protect our rivers."

Trout Unlimited and the Upper Colorado River Alliance continued to insist a free-flowing channel to bypass Windy Gap Reservoir was essential. Northern and Denver Water had more modest fixes in mind. Denver Water proposed aquatic habitat restoration work on the Fraser and Williams Fork Rivers. And Northern and Denver Water both proposed they monitor water temperature on the upper Colorado River, and possibly release water in late summer if temperatures rose high enough to threaten trout. "This seemed like a bunch of toothless talk," Bud says.

Other river advocates joined Bud in his skepticism. Mely Whiting, counsel for TU, commented, "The present mitigation plan doesn't get the job done." And the concerns of river spokespeople like Bud, Tony, and Mely were shared by the wildlife commissioners responsible for deciding the adequacy of the mitigation proposals. According to a *Denver Post* story by Scott Willoughby, after the commissioners heard Northern and Denver Water present their proposals, they "questioned whether additional protections might be needed to guard against high water temperatures and whether flushing flows contemplated by the plans would be enough to maintain channel health in the river."

According to Bud and Tony, some wildlife commissioners confided in them that they were sympathetic to their cause. The commissioners realized the Colorado River was damaged from past projects and diversions, and they understood dramatic actions were necessary to restore it to health. They had been given some data from Barry's study but not the report in its entirety. They knew they should have access to Barry's full report so they could base their decision on the best science.

It seemed Bud, Tony, and Barry were finally getting the traction they needed to focus attention on the upper Colorado's plight and build a bypass around Windy Gap. But things are seldom straightforward in the world of western water.

The commissioners unanimously voted to approve the mitigation plan Northern and Denver Water proposed—a plan Bud felt "did little more than pay lip service to the health of the river." When the Colorado Wildlife Commission made its final vote at a meeting in Grand Junction, Dave Buchanan reported in the *Daily Sentinel*:

> The commission's reluctance and frustration was obvious Thursday. "We can't stop the projects," said wildlife commission chairman Tim Glenn of Salida. "We're not totally comfortable (with the mitigation and enhancement plans) but what we have here is better than if we did nothing."

Better than nothing? Is this the best we can do for a river that has done so much for us?

WHEN MIKE KING MANAGES to make time for me in his hectic schedule and we sit down together, he explains he did the best he could to come up with an effective mitigation plan for the Colorado River headwaters. Fixing me with his steady blue-eyed stare, he insists he had to operate within the legal authority provided under state statute 122.2. My eyes glaze over. If you're having trouble falling asleep, and reading Senate Document 80 has failed to induce slumber, give state statute 122.2 a look:

> It is the intent of the general assembly that fish and wildlife resources that are affected by the construction, operation, or maintenance of water diversion, delivery, or storage facilities should be mitigated to the extent, and in a manner, that is economically reasonable and maintains a balance between the development of the state's water resources and the protection of the state's fish and wildlife resources.

I wake back up when Mike mentions trout. An avid fly fisherman, he grew up on Colorado's West Slope. He speaks wistfully of the days when giant stoneflies and green drakes filled the region's rivers and streams—in this sense he sounds a lot like Bud. Mike worries about Colorado's future, when a doubled population will put increasing pressure on resources, and citizens who are now children will have to protect what remains of the state's healthy watercourses.[22]

Conservation efforts must include youth outreach, Mike explains. If we preserve the Colorado River headwaters, but the next generation is too busy playing with their smartphones to care and they have no passion for rivers, the battle has been lost. Initiatives Mike spearheaded have sent underserved youth out of cities to spend time surrounded by mountain streams. In this regard, he is at the leading edge of smart stewardship of our state's water resources.

If, however, we help instill in youth a love of the natural world while handing off to them a dying Colorado River, we will have failed as stewards of one of our nation's most vital resources. One thing is clear to me after meeting with Mike King: I don't envy his position—in the middle of an endless

---

22   Colorado's population is projected to double by 2050, and the state's water resources are predicted to decrease due to climate change—creating a dangerous gap between water supply and demand.

tug-of-war between the West Slope and the Front Range, between development and preservation, between society's immediate water needs and the river's future health.

~~~~~~

Bud says he doesn't blame the DNR or Northern for trying to make Barry's report disappear—it was in their interest to do so. He says of Front Range water providers, "What we have here are some very professional, very astute public utilities people."

Nor does Barry blame Bud for what Bud did, in his desperation to save the river, when he got hold of a draft copy of the report.

Trout Unlimited, along with other organizations such as the Sierra Club and the Colorado Wildlife Federation, knew of Barry's report and were clamoring for its release. Trout Unlimited and the Upper Colorado River Alliance met with officials from the regional branch of the EPA. The EPA officials were, according to Tony, "totally appalled at the process." They agreed that damage caused to the river by past projects should be considered when assessing the mitigation plans for proposed projects. The EPA put pressure on Northern to start talking with TU, the Upper Colorado River Alliance, and Grand County.

"Northern didn't realize how deep the pockets of the groups opposing them were," Tony explains. When the EPA came knocking on Northern's door complaining about how the organization was handling public concern over the upper Colorado, Northern understood the permitting process could be held up for years. They knew they needed to negotiate, not obfuscate.

The message Tony and Bud sent Northern was simple: Get rid of Windy Gap Dam and your problems will go away; bypass the reservoir and you'll get your permit. But to strengthen their case, they needed Barry's report released.

Barry had told Bud about his research. Bud got the gist of

the report: The upper Colorado was a mess and serious efforts were necessary to repair it, including reconnecting the river around Windy Gap Reservoir. Bud was anxious to read the report, and he was desperate for it to be included in the conversation about the proposed mitigation plans.

With the deadline for the public comment period of the Windy Gap Firming Project EIS nearing, Bud found himself in Montrose, where Barry lives, after finishing a wilderness horseback trip. Before Bud returned to Denver, he convinced Barry to give him a draft copy of the report. Barry says, "I told Bud that the draft copy was for his eyes only—no copying and no distribution to anyone."

Bud respected Barry's request for discretion, but he believed that for justice to prevail, the report had to be made public. He understood there might be some backlash for Barry if the report was leaked, but he hated to see the hard work of a man dedicated to doing good stream science suppressed. "It wasn't fair to Barry," Bud says, "and it wasn't fair to the river."

Bud decided to give the draft copy to his attorneys. When the DNR continued to drag its feet on releasing the report, Bud's attorneys sent the draft copy to the U.S. Department of the Interior in Washington, D.C. The DOI dispatched to Colorado a representative who made clear that the way the state was handling the mitigation process was unacceptable. As pressure from the feds ratcheted up, Barry's report was finally released. And this groundbreaking scientific study that had remained hidden was revealed in all its telling detail.

～～～～～

BARRY'S REPORT CORROBORATED the message Bud and Tony had been spreading: The river was in dire straits due to dams and diversions, and the mitigation efforts proposed by Northern and Denver Water were inadequate to heal existing damage, much less offset impacts caused by increased diversions.

As the EPA came down hard on the state of Colorado for its reluctance to address the river's failing health, the media described this mounting criticism. In February 2012, journalist Bruce Finley reported in *The Denver Post*: "Federal authorities say a long-planned project to divert more western Colorado water to growing Front Range suburbs may cause 'significant degradation' of already deteriorating ecosystems along the upper Colorado River."

The EPA noted the final Windy Gap Firming Project EIS "in several sections, appears to state conclusions based on analyses or interpretation of results using methods that are inconsistent with scientific protocol."

It's worth lingering a moment over the phrase "inconsistent with scientific protocol." In a nutshell, the EPA said the state's mitigation plan was complete bullshit—what Bud and many other river advocates had been saying all along.

Among the EPA's censures was, not surprisingly, the exclusion of Barry's research from the final EIS. A document so long it could take months to read, the final EIS failed to mention the disappearance of all native sculpin and 38 percent of native aquatic insect species below Windy Gap Dam, as documented in Barry's study. Oops.... How this omission came to pass continues to spark conspiratorial speculation among veterans of the Colorado River wars. Though not the script for *Chinatown*, the story has enough intrigue to keep one's imagination busy.

This wasn't the first time the EPA had stepped into a scuffle over the upper Colorado and slapped some wrists. A few years previous, the Bureau of Reclamation's draft EIS of the Windy Gap Firming Project had provoked a flood of critical public comments; it had also incited the EPA to criticize the document for not adequately addressing the cumulative impacts of Northern's Windy Gap Firming Project and a proposed project of Denver Water. The EPA told the plain truth: The two projects affect the same watershed, so trying to separate their impacts makes no sense. A high school biology student could understand this. The Bureau of Reclamation, apparently, could not. Nor could the state of Colorado, which has had an uneasy relationship with science when it comes to the Colorado River.

State officials tasked Barry with studying the river's health; then they disregarded the results. Ken Kehmeier, an accomplished aquatic biologist with Colorado Parks and Wildlife, was at the center of the controversial wildlife commission hearings. Though reluctant to speak with me on record about the politics behind the state's mitigation plan, Ken stated emphatically that Barry's science was "absolutely solid." Ken told me he had been pressured to challenge the report but had refused to do so. He wouldn't reveal who exerted pressure on him to attack Barry's research. His guardedness is understandable. Ken has his job to protect, his career to watch out for.

Ken strikes me as a good scientist and a good person who has been drawn into political battles he would rather avoid. Like Bud, Tony, and Barry, Ken sees in wild trout the preservation of the world. But the world is tangled with political agendas, and a career in fisheries biology often demands one work not only in hatcheries but also in the halls of power where policies are made.

No one I interviewed could cite a flaw in the data or design of the study Barry led. Some of them grumbled, "Not everyone agrees with that study."

And not everyone agrees the earth is round. The inconvenience that the research causes powerful interests doesn't make the findings untrue. As astrophysicist Neil deGrasse Tyson says, "The good thing about science is that it's true whether or not you believe in it."

No scientist has offered an alternative explanation to Barry's conclusion that Windy Gap Dam and depleted river flows triggered a decline in the upper Colorado's health, as evidenced by the disappearance of key indicator species: the giant stonefly and the mottled sculpin. The research Barry led has withstood scientific scrutiny. Surviving political opposition has proved more difficult. Policymakers in Colorado have treated settled stream science rather like politicians at the national level denying evolution and climate change.

In 2008 the EPA stated the cumulative effects of proposed projects in the upper Colorado River watershed "could be severe, with irreparable harm done." Four years later, in 2012, even after Barry's report had been made public and had shown that the upper Colorado ecosystem was on the verge of collapse, business as usual in Colorado continued as its namesake river was sold down the river.

So great is the power of a thirsty metropolis it can move rivers across mountains. But no matter how much political will is expended by powerbrokers in cities, they cannot alter the truth. And the truth, once it has been revealed by science, has a way of finding the light of day, regardless of efforts to hide it.

WHEN WORD CIRCULATED WITHIN the DOW that someone had leaked Barry's report, Barry's boss sent him emails asking to whom he'd given copies. Barry took a lot of flak, but after turning in the report, he had retired and become a temporary employee. "I wasn't too worried about the backlash," he says. I get the sense that his bout of recklessness in passing a draft copy of the report to Bud had been prompted by his frustration with his superiors for suppressing science. He knew his research was sound. The study the DOW tasked him with was the culmination of professional achievements that spanned four decades. Barry had always aspired to be meticulous in his research. Scientists had not questioned his methods; researchers had not challenged his data. His upper Colorado River study was the capstone of a distinguished—and selfless—career.

When Barry was in the trenches of studying whirling disease, he didn't wet a line for ten years: He was too busy researching how to save wild trout to indulge in fishing for them. During the decade-long whirling disease crisis he was away from home at least half of every year. To accumulate data that would solve the riddle of the river's sickness, for three years Barry and his crew identified and counted insects in whipping snowstorms and punishing sun, in icy waters and searing heat.

David Nickum, executive director of Colorado Trout Unlimited, points out that Barry not only deserves credit for his work on whirling disease and his studies of the upper Colorado River's health, but also deserves gratitude for his work reducing bag limits to protect wild trout populations, and for his research determining minimum instream flows to maintain healthy fisheries in the state's rivers. Barry has become a hero to people who care about wild trout.

Barry and his team did their job and turned in their report on the upper Colorado River. The state did its job when forced to by the U.S. Department of the Interior, which pressured state officials to do the right thing—the right thing for the river, and for the people of Colorado.

IN THE SPRING OF 2012, journalist Scott Willoughby wrote in *The Denver Post* that proposed transmountain diversion projects in the upper Colorado Basin were "raising concerns over stress on an ecosystem considered at a tipping point since the disappearance of significant fish food sources such as giant stoneflies and sculpin, along with a corresponding decline in trout."

Bud responded to Willoughby's article in a letter to the editor, adding his own voice to a chorus of concern in the pages of *The Denver Post* and circulating throughout the state:

> Kudos to Scott Willoughby for his article on the Colorado River. If the mission of the Colorado Department of Natural Resources is to preserve and maintain Colorado's natural resources, then why is the agency the opposite side of the Environmental Protection Agency, Trout Unlimited, the landowners, Grand County, the Colorado River District and others on the solution to the continued death of the upper Colorado River?
>
> It is time for the governor to take action to protect the river before supporting more diversions while we watch it die. There is an opportunity to create a solution

that benefits all of Colorado, including the river, but the
governor needs to be part of the solution rather than part
of the problem.

Kudos also go to the scientists at EPA for recognizing
and describing this problem.

An oil and gas guy who in the past battled the EPA, Bud
cannot abide the river's deteriorating health. Nor can he tolerate reports of biologists being blatantly ignored. He says,
"No matter how many blankets you throw on a corpse, you
can't bring it back to life." Ignoring the findings of scientists
by burying their studies beneath bureaucratic procedures protected people's careers, but it has done nothing to protect the
Colorado.

"It's criminal what you're doing to the river," Bud said at one
of the state wildlife commission meetings, his voice cracking
with emotion.

Once, while standing atop Windy Gap Dam with an engineer from Northern, Bud looked downstream at the tainted
flow of the mud-bottomed river that had once run over clean
cobbles, and he said, "I just want to see the rocks in my river.
And the bugs. When I can see the rocks and bugs, I'll finally
know that everything is back to the way it's supposed to be."

Yvon Chouinard, founder of outdoor retail giant Patagonia, financed *DamNation,* a film that promotes dam removal
to restore the health of rivers. Chouinard said in an interview
with the *Los Angeles Times* he hopes the film builds on what
children are taught: "If you make a mess, you clean it up. You
don't just walk away from it."

That's how straightforward the situation with the upper
Colorado is to Bud. Northern and Denver Water broke the river; they should fix what they broke. "What I don't want is a
bunch of talk," says Bud. He is leery of "typical political agreements—which means do nothing now and hope that people
forget about the problem five years down the road." Proposals
to restore the river have amounted to nothing more than plat-
itudes and empty promises. Politicians make him sneer. "They

dance from one foot to the other so fast," he says, "it's like the best little whorehouse in Texas."

In conversations, when Bud isn't telling a story, he is moving the exchange toward the main topics he wants to cover. His email messages seldom contain an opening or closing—just the bullet points of the information he needs to convey. He doesn't want to listen to a politician or bureaucrat blather about how important the river is. He doesn't want to hear vague plans to remedy the problem. He wants to see the resource he loves restored to health. "I want to leave the river better than I found it," he says. "That's all I'm trying to do. This is a solvable injustice."

The stretch of river Bud is fighting to save is connected to other rivers and streams, many of which have had injustices of their own inflicted on them. Perhaps no other river in the upper Colorado watershed has been as badly abused as the Fraser.

Part IV

Chapter 14
THE WESTERN WHITE HOUSE

The Fraser River is important because it sends snowmelt from the Continental Divide toward the Colorado. But the Fraser is also a magnificent river in its own right, a serpentine stream that seduced President Dwight Eisenhower and countless others with its cold-water charms.

When President Eisenhower wasn't busy serving as Supreme Allied Commander during World War II, Commander in Chief of NATO, and President of the United States, he could be found wading in the waters of Grand County. At his "Western White House" he divided his time between conducting official business and casting for trout as he stood in the current of streams. Browns and bows, brookies and cutts were all brought to his net beneath the curve of his fly rod. Between fishing sessions, several bills were signed into law in Grand County.

President Eisenhower's wife Mamie was from Denver, a city he frequented. The Fraser River drew him into the mountains, where he stayed in a two-bedroom log cabin amid the magnificence of the upper Colorado River watershed. Grand

County residents sat around potbellied stoves in winter spinning tales of Ike, notorious for slipping away from his Secret Service agents to find solitude on streams. Locals often came across this friendly outdoorsman as he walked through fields or cast to rising fish—or wiggled his way through barbwire. One story spun around warm stoves in the chill of winter was told by a man who'd seen someone stuck in a fence with a fly rod in hand. As this concerned citizen had pried apart barbwire strands, he'd realized he was freeing the President of the United States.

The thirty-fourth president's favorite places to fish were the Fraser River and one of its tributaries, St. Louis Creek. When I was a teen and my family made a summer trip to Winter Park, St. Louis Creek and the Fraser River were my two favorite places to fish. Though back then, if I had read anything about Eisenhower and history I would have fallen soundly asleep. Now I'm wide awake to study Denver Water's history so I can understand why the Fraser and its tributaries are so radically diminished from when I was a boy.

Even when President Eisenhower was wading in the Fraser, its flows were already being reduced by Denver's diversion system beneath the Continental Divide.

~~~~~~~~

IN THE RAMSHACKLE OUTPOST of Denver, pigs wallowed in water ditches, causing cholera outbreaks. In 1872 the Denver City Water Company constructed a well, a steam pump, and four miles of mains to send water to homes. But sewage contaminated the system, spreading typhoid through the city. Throughout the 1870s and '80s, ten companies battled to control Denver's chaotic water industry. In 1894, local business leaders David Moffat and Walter Cheesman merged several companies to create a stable system for water treatment and delivery. But as population went ballistic in this prairie boomtown, water shortages loomed.

Supplying safe and abundant water to the ever-growing metropolis was deemed too important to be left to the vagaries of the free market. In 1918, Denver residents voted to create the Denver Water Board as part of the municipal government, preventing private companies from controlling the city's water supply.

In 1936, during the Dust Bowl and the Great Depression, the federal government financed an aqueduct beneath the Continental Divide, bringing relief to the thirsty and nearly bankrupt city. The Moffat Tunnel forced the Fraser to flow not west according to nature's design but east, toward money. This diversion allowed Denver's urban core to expand, until the water demands caused by a post–World War II population boom sent Denver on an infrastructure-building rampage.

In 1962, workers blasted the Roberts Tunnel more than twenty-three miles through the Continental Divide. One of the longest water tunnels in the world, this engineering feat allowed Denver to drink from Dillon Reservoir, which stored Blue River water behind Dillon Dam, allowing the city to slurp the river through the straw of the tunnel. As Denver drank, West Slope rivers shrank.

The largest municipal water utility in Colorado pursued its designs to divert more of the Colorado River headwaters— until the 1990 EPA veto of Two Forks Dam abruptly ended Denver's West Slope water binge. Following the veto, a former president of the Western Colorado Congress said, "It's been a nasty situation on the West Slope in the past. Whenever Denver Water wanted water they would just come and get it.... Now it's a new era for Colorado."

Two Forks was a public-relations Waterloo for the once triumphant water utility. After being annihilated by Trout Unlimited and other organizations in the battle to win the hearts and minds of Coloradans, Denver Water mounted a campaign to rebrand itself. The organization removed the image of a dam from its logo, hired a former Environmental Defense Fund attorney as chief counsel, and installed water meters at Denver residences. Engineers insist you can't manage what you don't measure; in 1992 Denver Water finished installing water meters throughout the city so it could quantify its water savings.

Director Hamlet J. "Chips" Barry III took over the reins of Denver Water and led the utility down a path paved with conservation initiatives. Spurred on by Barry's enlightened leadership, Denver Water started working with the West Slope on the depleted flows in the upper Colorado River Basin. And everyone lived happily ever after. No, not really.

SIXTY PERCENT OF THE Fraser's native annual flow is already diverted to Denver, yet the city and its suburbs are planning to swallow more of the river.

Denver's determination to further drain the Fraser began with the drought of 2002, when Colorado suffered through the driest year since recordkeeping began in the state more than a century ago.[23] Irrigation ditches sat empty in the baking sun, damaging Colorado's agriculture industry. When word spread about the spotty snowpack and shrunken rivers, visitors didn't show up with their dollars, harming the state's tourism trade. Lawns on the Front Range withered.

The 2002 drought shook the state from its water stupor: The epic dryness awakened Coloradans to the challenge of keeping civilization from collapsing in a water-stingy land. Emptied rivers and drained reservoirs drove Denver Water to shore up its supply to be ready for the next drought, and for the coming population surge. Couple our lust for growth with our primal fear of drought, and delicate dreams of environmental stewardship are trampled beneath the mad rush to build more water projects.

Denver Water proposed the Moffat Collection System Project in 2003 in response to the drought, which the organization claims pushed the city and its suburbs close to running out of water. Denver Water's critics contend that had the drought continued, lawns would have browned throughout the me-

---

23   The 2002 drought was surpassed in severity by the drought of 2012. According to state climatologist Nolan Doesken, the severe conditions of 2012 will likely be the new normal in the decades ahead.

tropolis, but no one was in danger of having the water pipes in their homes run dry. Nonetheless, Denver Water applied to the U.S. Army Corps of Engineers for a permit to expand Gross Reservoir to store more water diverted from the Fraser River system. A dozen years later, Denver Water's Moffat Project is still mired in controversy—much like Northern's Windy Gap Firming Project.

Denver Water contends this new supply would help ease a projected shortfall for its customers in coming years. According to the utility, additional flow from the Fraser River would also help balance its system and make it more flexible. Most of Denver Water's supply comes from its southern collection system: In yet another Continental Divide–spanning contrivance, water is diverted from the Blue River in Summit County through the near-marathon-length Roberts Tunnel into a chain of reservoirs along the South Platte River. Denver Water insists it must increase its diversions from the Fraser River to bolster its northern system—making the water supply the organization shepherds more resilient to droughts caused by climate change, and less vulnerable to extreme events such as forest fires and floods. According to Denver Water, storing more of the Fraser's flow would serve as a bulwark against disaster if a massive fire scorched a watershed in the southern part of its system—as happened with the 2002 Hayman Fire southwest of Denver. This conflagration clogged reservoirs with ash and sediment that washed down from blackened slopes, threatening the ability of Denver Water to provide a clean, safe supply.

To store additional water siphoned from the Fraser, Denver Water proposed raising the height of Gross Reservoir Dam, located in the foothills of Boulder County. This concept dates back to the Two Forks controversy, when environmental groups supported expanding Gross Reservoir Dam as an alternative to building a new skyscraping structure. Raising Gross Dam 131 feet would nearly triple the reservoir's capacity, allowing Denver Water to store more water from the already heavily depleted Fraser: At least 75 percent of the river's native flow would be rerouted across the Continental Divide.

Because the plan called for expanding existing infrastructure rather than building a new dam to flood a pristine canyon, some green organizations such as Western Resource Advocates, an environmental law and policy center based in Boulder, didn't oppose it.[24] Other organizations such as The Environmental Group and Save the Colorado have held firm in their opposition to increasing the size of Gross Reservoir. They argue the water shortfall for Denver and its suburbs should be addressed by making more efficient use of the water already being diverted, not by tapping more flow from West Slope rivers and streams. These groups will not agree to one more drop being moved across the Great Divide.

Defenders of the Fraser are quick to argue we are killing a river to grow green lawns for three months in the summer. Roughly half the treated drinking water used by residences in the Denver metro area is sprinkled on outdoor landscaping. People living on the Front Range, many of them transplants from moister climes, have replumbed the state's rivers to recreate the lush landscape of the East on the parched High Plains.

Historic photos of the Front Range show vast stretches of treeless prairie interrupted by occasional cottonwood corridors along river courses. Carpets of Kentucky bluegrass didn't soften the harsh landscape. Now, lawns so large they take landscaping crews several hours to mow suck up much of the Fraser, to the detriment of life nourished by the river's native flow. Some of the water diverted to hay meadows and used for snowmaking in Grand County drains back into the river. By contrast, water sent to the Front Range to grow lawns is gone from the Fraser forever.[25]

Whatever damage the Denver metro area's increased diversions would do to the West Slope's waterscape, Gross Dam

---

24   Western Resource Advocates said it wouldn't oppose the project as long as Denver Water promoted conservation and agreed to adequately mitigate environmental impacts resulting from reduced flows in the Fraser River and its tributaries.

25   Marijuana is another grass competing for the state's limited water. Cannabis is cultivated in commercial operations along the Front Range—these "grow houses" slurp up large quantities of water.

expansion would, somewhat ironically, enhance fish habitat on the East Slope. As things stand now, Denver Water can use its water rights to draw down South Boulder Creek below Gross Reservoir in winter. I have seen the stream dewatered so severely some winters, fish suffocate in frozen puddles and their corpses litter the dry bed. If Gross Reservoir Dam is expanded, an "environmental pool" in the enlarged reservoir will be dedicated to maintaining a healthy streamflow in South Boulder Creek throughout the year.

South Boulder Creek flows by the backdoor of the condo in Boulder where I lived for a decade. I'd like to see more water in this stream, where I used to spot one of my favorite birds, the black-crowned night heron, and where I crawled through tunnels of willow on the banks to watch trout working in pools, the white of their mouths flashing as they ate insects that followed the current. But is removing three-quarters of the Fraser River's annual flow too high a price to pay for green grass in our neighborhoods, for streams full of fish in our backyards, for recharged wetlands busy with birds along the Front Range?

The ailing Fraser may be out of sight, but for those of us who spent some of the best days of our childhood immersed in its waters, the river is seldom out of mind. When I see emerald lawns in droughty summers on the Front Range, I remember the Fraser as it once was. And I think about its starved flow now choked with silt and weeds, its fish struggling in the heat of shallow pools, the birds on its banks fleeing desiccated shores.

***

FORGET ABOUT GREEN LAWNS on the Front Range, Denver Water's public relations department insists: The Moffat Project will prevent an existential threat to the utility's service area.

Will Denver dry up and blow away if more water isn't diverted from the Fraser? When the EPA quashed Two Forks Dam, Denver Water Board Chairman Monte Pascoe stated at a press conference, "Rejection of Two Forks will have a devastating impact on the Denver metropolitan area. The effects from something like this aren't felt in a day or a month, but we are now in a terrible dilemma." Mr. Pascoe made this statement in 1989. Apparently the developers breaking ground on the McDonalds and McMansions that have mushroomed in the metro area during the ensuing quarter-century didn't attend that press conference.

At this point in the story, it's worth setting aside for a moment Denver Water's overblown rhetoric and inappropriate hyperbole to consider what Denver Water does. It saves our lives each day.

Consider a "water outage." Most of us aren't familiar with the term. We have all experienced the annoyance of a power outage. And sometimes the water supply in our home is briefly interrupted due to maintenance issues. But imagine if one day you went to turn on the tap and no water came out. Imagine if water didn't flow from the faucet for several days. Or several weeks.

We will never experience a deadly water outage on the Front Range because people like Jim Lochhead, Denver Water's CEO and manager, are working hard to make sure this doesn't happen. Water managers have done such a superb job of providing clean, safe water for us to grow our gardens and our children, we tend to take for granted the staggering amount of effort this requires. And we forget this crucial fact: Our modern water supply system is in large part responsible for the incredible leaps in lifespan that separate us from a chronically dangerous past when diseases like typhoid and cholera killed countless people. The Grim Reaper that once lurked in disease-infested water pipes has been banished by the dams and diversions that provide a steady supply of fresh water, and by the energy-intensive facilities that treat raw water to make it safe for us to drink.

Water providers like Denver Water and Northern Water continue a revolution that reduced human suffering on a tremendous scale. To open your faucet and fill a glass with potable water is to benefit from an achievement as important as antibiotics, childhood vaccinations, and other major medical advances that allow human beings to flourish. This narrative used to strike me as water provider propaganda—until I helped villagers in Uganda haul jugs of muddy water to children stricken with dysentery.

So, before we bash water providers for building dams and draining rivers, first we should sincerely thank them for saving our lives, and for providing one of the mainstays of modern civilization. Then we must do the hard work of studying proposed projects that affect the health of our watersheds. Just as doctors can be well-meaning but damage the health of patients, so can water professionals with good intentions kill a river. And it is up to us, the public, to hold them accountable for their mistakes and to take them to task for their misguided schemes. Along with Barry, Tony, and Bud, a stonemason in Grand County has, for many years, been doing exactly that.

Chapter 15
THE STONEMASON

Kirk Klancke retired a week before I met him, but he was working as hard as ever: Saving the Fraser is a full-time job. Before Kirk took me on a tour of the Fraser River Valley, we stopped by the Winter Park Ranch Water and Sanitation District office for a lunch of bratwurst and vegan hot dogs with Kirk's former employees. They thanked Kirk for "handing off a department in excellent shape." Kirk showed me around the Fraser Valley; then he headed off to meet U.S. Senator Michael Bennet to give the senator the same tour he'd just given me. Kirk has conducted this tour so many times and with so many people over the years he could lead it in his sleep. Yet each time he talks about the Fraser, his voice trembles.

Within sight of the ranch where the leader of the free world, President Eisenhower, chose to come when he could have been vacationing anywhere, Kirk sweeps his arm across an arc of mountains that surround the valley. This panorama provides a striking illustration of the scope of Denver Water's diversions. Kirk explains that from these snowy summits some thirty reaches of stream in the Fraser River watershed are tapped.

All but four of these siphoned streams lack legal protection from being dried up by Denver's diversions. "It's a system designed to suck the life from this valley," Kirk says, shaking his head. In some years the Fraser is skinnied down to a flow that would fit through a firehose. "Changing the state's water laws is basically impossible," Kirk notes. "Legislation is too slow, and litigation is too expensive. Education is the best chance we have to keep the river alive."

BEFORE KIRK SERVED AS a water and sanitation district manager, he worked as a stonemason. He learned the trade in Grand County when he moved to the mountains after quitting city life. He craved room where he could roam and surround himself with nature, as he'd done as a boy in Minnesota. He says, "When I was a kid, I'd make myself a bowl of cereal for breakfast and then go outside and disappear into the North Woods for the rest of the day."

When Kirk was a teen, his family moved to Kansas City, severing his connections with nature. He sought solitude in the weight room and absorbed himself in sports. Now in his early sixties, Kirk has a square-jawed face reminiscent of Arnold Schwarzenegger's chiseled visage, and Kirk's physique, sculpted in the gym and on the football field, is that of a bodybuilder. A football scholarship sent Kirk to college in Pueblo, Colorado. Before heading to the "Pittsburgh of the West" to play ball, he knew nothing of this gritty steel city. "It was a tough town to like," Kirk says in his succinct way. Fed up with cities, he dropped out of college and fled to Grand County. "Humanity disappointed me. People just weren't nice to each other."

In the Valley of the Fraser, silence spread like a silver blanket across the winter hills. Pines rose straight into the night sky, bright with stars and streaked with moonlight. Coyotes filled the quiet with weird yips and strange cries. No traffic sounds tainted the icy stillness that crept over the valley when the sun sank behind mountain walls. Kirk knew he'd found his home.

Fraser, once called "the icebox of the nation," lost the right to trademark this title to a town in Minnesota. "The harsh winters here help people be nice to each other," Kirk explains. He says when he first moved to the mountains he witnessed water mains freeze solid in winter; they didn't thaw until July. "But neighbors would run hoses to your house to make sure you had water without you even asking for help."

Many newcomers to Grand County tire of the cold and quit the area after a year or two. Kirk, who arrived more than forty years ago, is now a local. He says, "Everyone in Colorado came from somewhere else at some point in their personal or family history. When we give back to our communities, that's when we become true locals and real Coloradans."

Kirk arrived in 1971 as part of a wave of hippies fleeing cities and seeking meaning in the mountains. Grand County loggers were leery of these long-haired refugees from civilization who partied hard. But many denizens of Grand County had wildness in them. Kirk says he met a "wild pot-smoking cowboy" who taught him Nordic skiing. High in timbered hills where elk outnumbered people, Kirk cross-country skied from his front door. And he learned the trade of a stonemason, giving him blue-collar cred with local laborers.

A German artisan named Hans took Kirk on as an apprentice, teaching him the Old World craft of creating stone structures through careful selection and placement, coaxing and nudging each stone into place by feel and by practice—instead of relying on big gobs of cement to bind them together. "[T]rue masonry is not held together by cement but...by the warp of the world," wrote Cormac McCarthy in his play *The Stonemason*.

When Hans returned to Europe, Kirk purchased his business and continued his tradition of craftsmanship in Grand County. Kirk maintained his football physique by lifting heavy rocks each day. "I made a bad hippie because I worked so hard," he says. Kirk and his wife, Marianne, befriended three elderly women who lived in a cabin with no running water. Kirk hauled firewood for the women and helped them with chores. He was rewarded with stories from the valley's past, when Scandinavian loggers populated the hills. He learned how a prisoner of war camp in Fraser that incarcerated captured Germans during World War II had become part of the valley's lore; a man named Morris who ran the camp found wood for the prisoners so they could make instruments, and they formed a band that played for the locals.

One day after Kirk had made his way into the valley as a hitchhiking hippie, a man pulled his truck to the side of the road, gravel spraying as he swerved to a stop. On the truck's bench seat was a fat sack of money. Next to the money sat Morris, the man who'd run the prisoner of war camp. Kirk learned Morris had grown so close to the prisoners, he'd stayed in touch with them throughout their lives. Now Morris was working for the ski hill, and each day he drove the resort's cash to the bank—no armored car in this small town. Just when Kirk had been about to give up on humanity, a man with a moneybag stopped to pick him up. Kirk wondered what he could do to repay the kindness he'd been shown by community members.

Marianne was an artist and a strong advocate of volunteerism who inspired Kirk with her projects. Teens vandalizing homes and businesses had become a Halloween ritual around Fraser; as a form of redirection, Marianne spearheaded an effort to create a haunted house. She let teens take ownership of the project and organize it. "Vandalism on Halloween decreased," Kirk explains in an admiring tone. Marianne was his hero. She died a few years ago, one of many losses Kirk has endured in the Fraser Valley. But Marianne's legacy lives on in Kirk, who cannot say no to any community project he is asked to lead. He is notorious for double-booking and triple-booking his schedule. His new wife, Darlene, calls Kirk's routine "the rat race of retirement." Along with hauling the Lion's Club pancake trailer to fundraisers, he serves as president of the local cemetery association. He once had to excuse himself from visitors in his home after being called in the dark of night to clear paths through deep snow to bury the dead. "You have a really weird life, Kirk," one of his guests said.

~~~~~~

KIRK'S AWAKENING TO THE death of the Fraser River came on a backpacking trip when he was twenty years old. While following Iron Creek, a tributary of the Fraser, down to the valley floor, the little stream burbled and sang until Kirk encountered a concrete wall—one of Denver Water's many diversion structures that siphon the Fraser River headwaters to the Front Range. Somewhere in the hidden distance, the city of the plains sparkled with green lawns, while downstream of the diversion, Iron Creek completely disappeared. "I saw the one hundred percent death of a river," Kirk says. He wondered how an organization could get away with killing a stream. "But I was too immature to do much about it," he says. "I took a piss in Denver's collection system before I finished my hike."

After a few years passed and Kirk's rambunctious mind matured, his concern for the Fraser, a river beaten to its knees by abuse, prompted him to defend what he loved. When he first started spreading the word about the Fraser, he was greeted with confused silence. Today, you would be hard-pressed to find a local in Grand County who doesn't know the Fraser River is in serious trouble. Kirk's brother, Dr. Kim Klancke, told me, "It's gone from Kirk complaining to a movement."

Like Kirk, Kim is a large man with a large heart. A cardiologist committed to helping suffering people, Kim tends to neglect his own diet, and his big frame carries an unhealthy amount of weight. The Klancke brothers are incapable of ignoring anyone or anything in need. For their kindness they often bear large burdens.

During his decades of spreading the word about the Fraser's plight, Kirk has been on the receiving end of both recognition and wrath. *Field and Stream* magazine named him one of its Heroes of Conservation in 2011. Kirk bought a suit for the first time in his life and headed to Washington, D.C., to receive the award—and to lobby members of Congress about the Fraser. U.S. Senator Mark Udall gave a speech on the floor of Congress honoring Kirk's advocacy work. "It made my mom proud," says Kirk, a big grin breaking apart his modest demeanor. In 2014, *National Geographic* published an article praising Kirk's efforts as a steady voice for the Fraser and a smart broker of deals—a determined guy who manages to get things done to save the river.

Perhaps the most telling measure of Kirk's impact is not prestigious awards and national media coverage but that every person I spoke with who lives in Grand County knows who Kirk is and what he does. The name "Kirk Klancke" is synonymous with "Save the Fraser"—a phrase that adorns stickers and beer mugs and wineglasses from Winter Park to Granby. It has become a brand that identifies locals when it marks the bumpers of their vehicles. Kirk laughs and says, "Some of them aren't sure what they're saving the Fraser from. But it's a start."

Kirk didn't earn a college diploma, but he knows as much about Colorado water law as an attorney. He figured out how to navigate Colorado's legal labyrinth that allows the Fraser's tributaries to be dried up, and the loopholes for leaving water instream, beginning with his tenure on the East Grand County Water Quality Board. After learning to talk in cfs and acre-feet, Kirk was appointed board chairman because of his self-taught water fluency. Fed up with watching the Fraser suffer, he wrote an essay about damage to the river caused by Denver Water's diversions. This essay got the attention of American Rivers, which in 2005 named the Fraser the third most endangered river in the nation. And this listing got the attention of Denver Water.

Kirk says, "A Denver Water employee called to tell me that if I didn't get the Fraser River taken off that list, they'd come after my personal assets." Kirk was shaken by the threat. He feared he'd lose the life he'd worked so hard to construct. He likes to say he doesn't have a million dollars but he has a millionaire's lifestyle. The house he built, sited on land sold to him for a song because of work he did for the owner, has one of the finest views in Grand County. Kirk had no idea how he would defend himself against Denver Water if they came after his home. "I was thinking that in modern-day Colorado we don't kill people and dump them in the river," he says, laughing. Then his brow knits and his eyes narrow. "But I was still pretty worried."

Kirk thinks Denver Water wasn't accustomed to being challenged. He says, "They had worked in a vacuum for so long they'd gotten used to getting what they wanted from the Fraser River." They didn't seem to know what to do with Kirk—other than threaten to sue him to keep him quiet. A soft-spoken stonemason had stirred the wrath of an organization with the will to move rivers across mountains. It took the passion of nonprofits, along with the wealth of multimillionaires, to keep Kirk's cause alive.

Chapter 16

THE HIGH COST OF SPEAKING TRUTH TO POWER

fter receiving that unsettling call from Denver Water,
Kirk phoned American Rivers to tell them about the
threat. "They told me I should call Denver Water and let
them know that the Fraser had just been moved up to number
two on the list. They said that their attorneys would protect
me. They welcomed the publicity for the river that a public
fight would bring."

And so Kirk forged ahead, telling the truth about the Fraser
and making enemies along the way. Kirk says he once reached
out his hand to Denver Water leader Chips Barry, who drew
his own hand back and said, "So *you're* the son of a bitch."
Kirk told Chips he was looking forward to working together.
But the most powerful man at one of the most powerful water
providers in the West refused to shake Kirk's hand and walked
away.

I can't verify this tale told by Kirk—Chips Barry died a
few years ago, and no one else witnessed the incident. But I've
spent enough time with Kirk on the river and in his home to
believe his morals are as sturdy as the stone walls he stacks.

Kirk's former employees speak of his honesty and integrity. His partners on river advocacy projects praise him effusively. I've heard tell of Kirk making an enemy or two in the Fraser Valley. But no one in a community so small it forces you to rub up against your neighbors when you gather in town fails to chafe a few people from the friction.

Chips Barry's many admirers have portrayed him as a charismatic peacemaker with a disarming sense of humor in the water wars across the Great Divide. They have also painted him as a conscientious steward of Colorado's natural resources who pointed Denver Water in the direction of environmental responsibility. By all accounts he helped soften the damage Denver had done draining the West Slope with disregard for the consequences to rivers and local communities. Chips Barry was a warm hug compared to Glenn Saunders, a cold slap in the face. As attorney for Denver Water, Saunders "attempted every means that come to mind in the past 30 years to steal western Colorado water," according to a Grand Junction newspaper.

Kirk's anecdote doesn't diminish the admirable efforts of Chips Barry to bridge the divide between Denver Water and the West Slope. It shows the man had a temper as well as a sense of humor—which is to say, he was human. Perhaps it also shows Denver Water wasn't as dramatically transformed after Two Forks as is often portrayed. Perhaps the organization is neither virtuous nor malevolent but, like the human heart, is forever in conflict with itself as it struggles to adapt to the demands of a changing world.

The Two Forks defeat wasn't the first time Denver Water was forced to do the right thing for rivers—that precedent was set long before Chips Barry's leadership. In 1978 Denver Water entered into something called the Foothills Agreement. This compromise ended lawsuits brought against the organization by environmental groups over a controversial proposal to build the Foothills Water Treatment Plant and Strontia Springs Dam. Opponents claimed these projects would not only lay the foundation for more water diversions from the West Slope but would also promote low-density urban sprawl and the excessive use of automobiles, leading to air pollution.

To secure the permitting for its project, Denver Water agreed to mitigate the dam's adverse effects on the South Platte River. It also agreed to implement a "credible and measurable water conservation program." The organization now touts its award-winning efforts to conserve water. It should be congratulated for these effective water-saving measures, to be sure. But the organization was *forced* to do the right thing. History shows Denver Water and other Front Range water providers can be shoved down the path of sound environmental stewardship. Is there any reason to believe they will walk that path willingly?

Arguably, Denver Water has displayed sound environmental stewardship only when threatened with the hammer of litigation. If it behooves Denver Water to negotiate rather than litigate, its leaders will negotiate. If they can avoid lawsuits by conserving water, they will adopt conservation as one of their organization's core principles. Based on a careful study of Denver Water's past, it seems reasonable to conclude that if people like Kirk Klancke were to stop acting as watchdogs on the West Slope, Denver Water would try to drain the Fraser of every last drop.

Perhaps Chips Barry was more of a pragmatist than an idealist, and he didn't so much change the culture of Denver Water as alter the way the organization does business when backed into a corner. Maybe he helped Denver Water make environmentally sound decisions because there were no other decisions the organization *could* make after the EPA's cancellation of Two Forks. Is it too cynical to conclude Denver Water will do the right thing for the river only after it has exhausted every other option? The truth is in their charter: This municipal utility is charged with providing water to people in its service area, not with protecting the health of the Fraser. That job fell on Kirk's shoulders, which, as providence would have it, are strong from years of carrying stones.

KIRK HELPED START A grassroots organization called Friends of the Fraser. "We had a brochure and a website and nothing else," he says with a shrug. "It was pretty amateur."

After Kirk's essay prompted American Rivers to list the Fraser as one of the nation's most endangered rivers, local papers reported the story. Word spread. Before the listing, few people in Colorado knew where the Fraser was located, let alone it was endangered. "It helps to have ink on your side," says Kirk.

Ken Neubecker, one of the state's most outspoken advocates for healthy rivers, came to Grand County as a representative of Trout Unlimited. "I thought maybe TU wanted to sue me," says Kirk. The Sierra Club had threatened a lawsuit if Kirk's water district didn't take measures to control water quality. Ken assured Kirk TU wasn't there to sue him—they were there to help. "It was David and Goliath," says Kirk. "TU gave us a slingshot."

Emboldened by TU's support, Kirk set out to expose what he saw as gaping holes in the environmental impact study funded by Denver Water for its Moffat Project. Denver Water's science was based on stream modelling. Kirk, who spends his days standing in streams instead of making computer models of them, explains that as the Fraser's depleted flow spreads across its native bed, the slow-moving, shallow river heats up swiftly in the sun. Denver Water, however, argued air temperature, not the river's decreased volume, was causing the stream temperature to rise. Kirk says, "I tell people it's real simple. If you take a glass of water and a gallon jug of water and put them in the sun, which one is going to heat up faster?"

Trout Unlimited gave thermometers to anglers and asked them to measure and record stream temperature from various places and times on the river. Then TU created charts based on this data. This effort at "citizen science"—training and equipping people without science degrees, but with a keen interest in the natural world, to collect data in the field—has a long and distinguished history (think Audubon bird counts). Citizen science has also enjoyed a recent spike in interest in academia. For example, University of Colorado ecologists studying how climate change affects pikas in the Rockies have enlisted hikers in the effort. Regardless, Denver Water dismissed TU's effort. "They laughed in our faces," says Kirk. "They were howling with laughter and saying, 'Fly fisherman with thermometers? That's not science!'"

Trout Unlimited volunteers recorded stream temperatures as high as 74 degrees—lethal for trout. Denver Water insisted this was impossible because their modelling proved it. Computer programs in air-conditioned offices calculated no problems with the river; therefore the river was fine. Case closed, as far as Denver Water was concerned.

The cost of professional data logging machines to record stream temperature was beyond the budget of Kirk and TU. Grand County, however, is home to a surprising number of the well-heeled. The area may not have the cachet of Aspen, but the Fraser River Valley attracts plenty of multimillionaires— some of whom rallied to Kirk's cause.

Kirk tells me one day a man named Bob Fanch dropped by his office and said he'd heard through the grapevine what Kirk was up against. He wanted Kirk to know he was willing to help.

A telecommunications pioneer, Bob Fanch also developed Devil's Thumb Ranch, named by *Travel and Leisure* magazine one of the top twenty sustainable resorts in the world— and it happened to be in a meadow across from Kirk's house. Kirk had once done a favor for Bob without knowing his background. Kirk had reflexively helped his neighbor—similar to how, many years ago, when Kirk was hitchhiking after arriving in the valley, the man with the moneybag on the seat of his truck skidded to a stop to give Kirk a ride. Camaraderie among community members, some of whom possessed the wealth of small nations, played a crucial role in reversing the Fraser's decline.

KIRK TOOK BOB FANCH up on his offer and brought him to a meeting at Denver Water headquarters. Kirk says Bob sat quietly and listened as Denver Water employees preached the superiority of their computer models over Kirk's "junk science." Bob explains that after the first meeting he attended with Denver Water, he realized "we needed our own real science based on field data versus Denver Water's models."

Like Bob Fanch, Roger Newton, another wealthy philanthropist, worried about the Fraser's fate and wanted to help. Roger footed the bill for temperature data logging machines through his Esperance Family Foundation. Studies using these data loggers were supervised by a professional science organization, which corroborated the findings of TU's citizen scientists. River temperatures were rising to levels lethal for trout, and low flows were the main cause of this increasing heat. Denver Water stopped laughing: A strong scientific case that their waterworks were killing the river quieted them, according to Kirk. The noose of their diversions was choking the life from the Fraser and could no longer be ignored.

Another battlefront in the science wars in which Kirk could declare victory was his metric of Denver Water diverting 60 percent of the river's native annual flow. Denver Water disagreed. By citing data from gauges far downstream on the river, they came up with a number half as large. Kirk explains that as people in the know pointed out the ridiculousness of this ruse, Denver Water was forced to back down. Now the figure of 60 percent is no longer disputed. Kirk's math prevailed. He had a stonemason's common sense and a mountain man's habit of telling the plain truth.

Though placing blame for the Fraser's decline squarely on Denver Water, Kirk concedes Grand County isn't a perfect steward of the river. Winter Park's diversions for snowmaking reduce the flow that the Fraser needs to flush itself of sediment.

Sanitation and stormwater challenges in Grand County's own backyard also affect water quality in the Fraser. But any reasonable person can see that transmountain diversions have sapped this river's strength.

Similar to how Kirk had realized rising temperatures in the river were a problem, he saw sandbars clogging channels and knew something was wrong. Sediment was not moving in the river's weakened flow. But science needed to be done to demonstrate a causal link between Denver's diversions and the deteriorating condition of the riverbed. And those studies were not cheap.

Bob Fanch's Sprout Foundation funded a study of sediment transport in the Fraser. A follow-up study was paid for by Fanch's foundation, and by Cathey McClain Finlon—yet another deep-pocketed donor drawn to Colorado by the stunning scenery. This Colorado Business Hall of Fame member built a local advertising company into one of the top advertising firms in the United States.

Because the best and brightest bring their talent to Colorado after being attracted by its mountain paradise allure, drying up the state's rivers is a monumentally bad decision. Colorado's business leaders should be as outraged as environmental activists over the failing health of the Fraser. Successful entrepreneurs can live and work anywhere in the world. They choose to base themselves in Colorado in large part because of healthy watersheds.

There's an old saying: Trout don't live in ugly places. They also don't live in ecologically damaged places. And tourists don't travel to ecologically damaged places to spend their dollars. Similar to how a healthy watershed maintains the ecosystems on which all life depends, a trout-filled river has much broader implications for the state's economy than just fishing.

According to TU, fishing in Colorado, when related spending is factored in, is a $1.3 billion-per-year industry. And the whole outdoor recreation economy, which generates $10 billion each year in Colorado, relies on healthy rivers. Even more important, healthy rivers are an integral part of Colorado's world-famous brand.

In a state where the quality of life catches the attention of people who can visit, live, and locate their businesses anywhere on the planet, a brand known across the globe for outdoor excellence is an invaluable asset for current citizens and future generations. Healthy rivers, not green suburban lawns, define Colorado. If our rivers die, we lose our competitive advantage over other states—and the best and brightest take their skills and their money elsewhere. "Come to life," the motto coined by Colorado's tourism office, will become sadly ironic if the state continues to kill its namesake river and storied streams like the Fraser.

The sediment transport studies paid for by wealthy philanthropists determined the Fraser's weakened flow couldn't flush the stream clean. Led by Dr. Brian Bledsoe, a widely respected environmental scientist, the research revealed coarse gravels were not being moved, and fine sediments were forming hard veneers on the streambed. This armoring effect smothered the spaces between gravels and stones essential for trout spawning and aquatic insect survival—similar to findings in Barry Nehring's report on the Colorado River below Windy Gap Dam. Dr. Bledsoe's report concluded that the increased diversions to the Front Range being proposed would further deprive the Fraser of the flushing flows needed to scour away algae and silt and rejuvenate the river's stagnant bed.

A solid body of science now supports what Kirk realized all those years ago when he stood atop a diversion structure and saw dry stones downstream. Simply put, when too much water is diverted from a river, it dies. A child could understand this. But municipal water utilities aren't run by children—they are run by people who need paychecks and pensions. Denver Water isn't funded by tax dollars—its revenue comes from charging developers for tap fees and customers for the water they use. A depleted Fraser River means money in the bank in Denver. It also means people with money will be less likely to come to Colorado.

Kirk points out that not only did wealthy residents of Grand County fund scientific studies of the Fraser, but also some of them spread the story of the imperiled river to people in positions of power. "It takes a village to save a river," Kirk says. He explains that three factors combined to put pressure on Denver Water to heal the river it had harmed: the awakening of citizens and politicians in Grand County to the source of their economy and way of life being sucked dry, the legal expertise and financial resources of TU, and the generosity and connections of moneyed supporters. David's slingshot had been augmented with an arsenal.

But the state's largest water utility still had the most powerful weapon of all: Colorado's system of water law, with its prior appropriation doctrine that allows entire rivers to be diverted from one basin to another. In theory, the instream flow rights in the Fraser and its tributaries protect streams in Grand County from being drained. But these instream flow rights are junior to Denver's rights—meaning in times of water scarcity, Denver gets to divert, and streams in Grand County run dry.

Kirk's only defense against the power of prior appropriation is education.

〜〜〜〜〜〜

KIRK KLANCKE IS A BOOK unto himself. Doing justice to his innovative efforts to inform the public of the Fraser River's ecological collapse is not possible in this limited space, but a few highlights follow.

As president of the Colorado River Headwaters chapter of Trout Unlimited, Kirk helped fund a billboard on Interstate 70 that admonished drivers:

DON'T SUCK
THE UPPER COLORADO RIVER DRY.

Kirk gave talks to the Lions Club and Rotary, where he made inroads with the forty-year-old-plus demographic. To reach people in their twenties and thirties, he threw a bash called Riverstock on the fortieth anniversary of Woodstock. For many years, Kirk has gone into classrooms to talk with young people about the river; more importantly, he has taken them out of classrooms.

On field trips in the Fraser watershed, kids learn the science of the river the way Kirk did—by participating in research. Kirk takes them above Denver Water's diversions so they can smell the mossy banks of healthy streams. They scoop nets into the current to identify and count many forms of life. They hear the water talk and hold it in their hands, feeling it slip through their fingers as it wiggles its way downriver, passing stones tattooed with the cocoons of caddisflies. Kirk also takes the kids below the diversions, where they walk the dry beds of reaches long dead. Schoolchildren learn on these field trips they will grow up in a state where it is legal to kill a stream, and the only way to prevent this is if they care enough to someday put a stop to it.

In *Tapped Out,* a short film produced by The Story Group for Trout Unlimited, Kirk asks people on the streets of downtown Denver where their water comes from. Nine out of ten tell Kirk something along the lines of "the sink" or "the sky." A few years back, before I set out to make sense of why the West's rivers are dying, my answer to Kirk's question would have been just as laughable. We are disconnected from so many things in our hyper-connected world. We can instant message someone in India, yet the source of our water, a substance we must ingest every few hours to stay alive, is a mystery to us. And for most people, the health of rivers is not a pressing issue.

But Kirk is a dogged optimist. He says, "When I was a boy, my family rolled down the windows of our car and threw our trash on the side of the road." What was once accepted practice is now seen as deplorable. Kirk believes a day is coming when killing the Colorado River headwaters will be viewed as an outrage. He envisions a time in the not-too-distant future when cultivating Kentucky bluegrass lawns on the semi-arid plains of the Front Range is considered as tacky as wearing a fur coat. He imagines East Slope cities making use of every option to increase their water supply in their own backyards: lining leaky canals, reusing wastewater, and tapping into the vast supply of outdoor landscaping water—a virtual reservoir larger than the amount of water Two Forks Dam would have stored. Kirk says, "I imagine a day when draining the rivers of the West Slope is an absolute last resort, not the first tactic of water providers on the Front Range."

Kirk's efforts at educating his community have paid off in spades. Take, for example, the public comment period of the draft environmental impact statement (EIS) for Denver Water's Moffat Project. This study has been scoffed at by scientists, environmental organizations, and the public ever since it was issued in 2009. Kirk says that of Grand County's 14,000 residents, 3,000 submitted comments, almost every one of them critical of the draft EIS. "I was proud of my community," says Kirk, a smile softening his stony jaw.

The U.S. Army Corps of Engineers, the agency tasked with issuing a permit for Denver Water to proceed with its project (what's referred to in river-geek jargon as a "Clean Water Act Section 404 Permit"), was overwhelmed by the volume of comments submitted. The deluge of feedback from the community prompted the Corps to slow down and study the project's potential impacts. The scientific field studies Kirk and TU initiated led the EPA and the Corps to scrutinize the dubious models Denver Water had submitted.

To say Kirk Klancke had become a thorn in the side of Denver Water is to be in danger of understatement. More like a bazooka blast in their side. But Kirk tends to avoid conflict. He'd rather go fishing or skiing than argue and fight. Like Bud, Kirk is just a guy who saw that something was wrong, and one day he spoke up, and he's been speaking up ever since. "It's gotten so much bigger than me," says Kirk. Now there are many dozens of Kirk Klanckes fighting for the Fraser. The original Kirk Klancke, the stonemason who started it all, is relieved to be retiring from river politics. "I'm ready to let the best science take over," says Kirk. "I'm looking forward to a future where proactive science-based solutions that improve the river's health replace all the fighting. I'm looking forward to spending less time arguing with Denver Water and more time on the river with my grandkids."

In Kirk's office hangs a framed piece of the Congressional record—the speech Senator Mark Udall gave honoring Kirk. Representative Udall stated that Kirk's work "embodies what I have long known to be true. We don't inherit the Earth from our parents—we borrow it from our children and the generations that follow."

People like Kirk are bringing national attention to a local issue of international importance. How the most powerful nation on Earth chooses to care for the Colorado River will set a precedent for developing countries across the planet. And the repercussions will spread far into the future. In rocks along the river's course you can trace ripples left by ancient seas. How we steward our rivers now will leave traces that pass through time into the stories of future worlds looking back on ours.

DURING THE FINAL COLORADO Wildlife Commission meeting discussed in the previous section, Kirk was approached by one of the commissioners, who pulled him aside before the meeting began and said, "You're not going to like what you hear today."

"But the meeting hasn't even started," Kirk replied. "The public hasn't spoken yet."

In places of power, decisions about the river had already been made. Kirk was so distraught after this meeting he fled to the desert to collect his thoughts. He considered leaving the state. "I came close to giving up on Colorado," he says. "I didn't think I could stand to stay and watch the Fraser die."

While traipsing through the desert in a funk, he stumbled upon a man from Idaho and told him his tale. The man laughed and said it was the same all over the West. He told Kirk that blind growth at the environment's expense—the kind of development that destroys the very reason people value places—was happening apace in Idaho. He warned Kirk he'd find the same thing in Montana or wherever else he tried to hide in the Rocky Mountains.

The Fraser Valley had provided Kirk with so much beauty and peace when he was a young man, his faith in humanity, shaken in the streets of cities, had been restored. He says, "I realized I needed to fight for the river—not for me but for my grandkids."

Kirk came back home to protect the waters that had once washed away his pain. His father had been killed when a plane he was flying in crumpled against the mountains that surround Grand County. Kirk's wife and one of Kirk's brothers both died young. One of Kirk's best friends, an athlete who played on the national rugby team and skied competitively, was stricken with a rare neurological disorder that left him partially paralyzed. When Kirk was overwhelmed by so much loss, so much inexplicable suffering, the one thing that had made sense to him was the river. Standing in the Fraser's current he had felt his cracked heart heal as the water flowed past him. The cruelty of this world is beyond comprehension. But so, too, is the beauty of a riffle that runs into a pool paved with river-smoothed stones.

Science is the most precise tool we have to make sense of how rivers function, but there is also deep mystery in their movements, surpassing our powers of description and understanding. The rivers of the world gathered their waters long before we arrived; those waters will flow long after we have departed. From a vantage point on a mountaintop above the Fraser Valley, the river loops like a dropped rope. And from the river's twists come unexpected turns of story.

At Kirk's first Riverstock concert, a man took the stage between songs and said, "We can save the Fraser. But first we have to stop hating each other." He was Tom Gougeon, a Denver Water Board member, and he got the biggest applause of the day. "I think Tom was sincere," Kirk tells me. And Kirk, though he laughs easily and often, is nothing if not sincere.

This craftsman with arms muscled from moving stones has been known to shed tears in public when speaking about the river that saved him.

~~~~~~~~

KIRK CHOKES UP WHEN he takes me on a tour of Denver Water's diversions. For mile after mile as we drive he points out ditches and pipes. We stop to peer through the grate of a giant metal box where a stream diverted from the forest falls downward into darkness. Signs shout at people peering at the source of their water: "DANGER Keep Away."

Kirk shows me Jim Creek, a tributary of the Fraser cut in half by a concrete wall. Above this diversion, trout chase mayflies in a brook that flows in a boisterous rush. Downstream of the barrier oozes a silent trickle of slime. While we stand atop this structure that drains the river of water and life, Kirk tells me he used to take his daughters to St. Louis Creek for picnics on its shores. Into icy pools they would jump, screaming and giggling as their lips turned purple and their limbs trembled before they fled the chilly stream to shiver themselves warm on its banks. Now, when Kirk returns with his grandkids to this spot, they wallow in the warm flow.

"When my daughter saw that kids were able to stay in the river so long without getting cold, she started crying." Kirk pauses and swallows hard. "She said to me, 'Dad what have they done to the river?'"

When Kirk and I return to his house that evening, I meet one of his grandkids, Ellen Koski. She tells me she likes to get away from the city and visit her grandfather in the mountains. "It's where some of my best memories are," Ellen says as she cooks s'mores over the snapping flames of a fire. Past stone walls built by Kirk, a stream swollen with spring flood surges through the valley, and sparks from the fire rise toward the stars.

Ellen recently won the Miss Teen Colorado competition. For the essay portion, she wrote about diversions to the Front

Range drying up the Fraser, and she encouraged people to water their lawns less.

"I brought Ellen to the river," Kirk says with a big grin. "But she figured it out on her own."

Chapter 17

HOLDING THE RIVER HOSTAGE

For several years Denver Water has been holding the Fraser River hostage while it negotiates taking even more of the Fraser's flow. Denver Water's website states, "Through enhancements that are contingent on the Moffat Project, the project will provide many benefits to the West Slope, making the West Slope better with the Moffat Project than without it." It's worth spending a few moments dissecting this strange logic.

The health of the Colorado River headwaters is declining mainly due to diversions to the Front Range. One of the organizations responsible for these diversions is offering to repair some of the damage—but only if allowed to divert more water. The looming death of the upper Colorado River is tragedy bordering on comedy. Gary Wockner, executive director of Save the Colorado, puts it bluntly: "You can't save a bleeding patient by cutting another artery."

State laws that don't require Denver Water to reverse the damage its diversions have caused evolved through mineral rushes and irrigated agriculture during the settlement of the

Colorado Territory, when wagon wheels were the leading edge of technology. A sensible water policy for the twenty-first century would, of course, place protecting the health of river ecosystems on equal footing with the need to divert rivers. Not in the state of Colorado in 2015. Denver Water will fix the river it has broken if—*and only if*—it is allowed to take more water.

Diverting more flow from an already depleted river while at the same time improving its health seems to subvert logic. Indeed, conversations about the Colorado can seem downright Orwellian. George Orwell's famous formulation in *1984*—"War is peace. Freedom is slavery. Ignorance is strength."—could become, in the weird doublespeak of the Colorado water world, "Diversions are improvements. Less is more. Death is life."

IN SPRING OF 2014, Jim Lochhead, Denver Water's CEO and manager, announced with considerable understatement, "After being in a permitting process for more than 10 years, we are pleased to see the release of the Final Environmental Impact Statement for Denver Water's Moffat Collection System Project." The final EIS, conducted through the U.S. Army Corps of Engineers, determined the project would help prevent a major Denver Water system failure that could leave metropolitan areas like Arvada and Westminster short of water.

Some environmental groups insist the amount of water the project would provide—18,000 acre-feet per year of firm yield—could be supplied by measures like making low-flow water appliances mandatory for residents within Denver Water's service area. But the story becomes nuanced: The organization pushing legislation requiring higher efficiency for water appliances sold in Colorado is none other than Denver Water.

Furthermore, the Moffat Project expands an existing piece of infrastructure rather than building a new dam—an approach in line with an environmentally responsible water ethos in the West. The proposed project has also involved more negotiations with groups advocating for protecting West Slope water resources than any previous transmountain project in the state's history. Many of those groups have praised the willingness of Denver Water to listen to their concerns and to come up with solutions to problems plaguing the upper Colorado River watershed.

Denver Water points out it uses 2 percent of the state's water to supply 25 percent of its population—and the Denver metro area creates much of the state's economic productivity. It's not surprising a utility would invoke utilitarianism, the greatest good for the greatest number, when defending its water diversions. Denver Water's supporters insist the animus directed at Kentucky bluegrass in the city would be better focused on grass grown by agriculture. Of the state's overall economic pie, farms and ranches generate a small sliver relative to the service sector, yet are responsible for 86 percent of the state's water use—much of which is squandered through flooding fields to grow thirsty, low-value crops like alfalfa.

Further complicating the state's grass wars, the water provider so many river advocates love to hate invented xeriscaping.[26] By using drought-tolerant plants appropriate to a dry climate, this landscaping innovation reduces wasteful water use in cities. Though the number of people Denver Water serves has risen in the past few decades, the utility's overall water use has decreased 20 percent due to conservation efforts.

Denver Water's claims to environmental virtue stretch all the way back to 1936, when the organization advertised on trolleys to encourage customers to save water; they extend to present initiatives like providing rebates to customers who install low-flow toilets. Even so, Denver Water's efforts to divert more flow from the Fraser enraged many West Slope citizens. A deluge of comments followed the 2009 release of the draft EIS for the proposed Moffat Project.

---

26   "Xeriscaping" is derived from *xeros*, the Greek word meaning "dry."

In the draft report, as thick as a Denver phonebook and about as interesting to the average person, the EPA and the Colorado Division of Wildlife identified the project's potential impacts on the upper Colorado River watershed. The comments in the report made clear the widespread support among the public for protecting the Fraser River. Individuals, environmental groups, and community representatives all weighed in on the need to restore and maintain the Fraser's health. Dark clouds of litigation billowed on the horizon. But they didn't build into a storm that flooded the state's courtrooms because Denver Water did the right thing. Dark skies cleared, and a bright new era dawned. Or so the story goes.

IN SPRING OF 2011, PEACE broke out in the West's water wars. The Colorado River Cooperative Agreement resulted from then–Denver Mayor Hickenlooper's efforts to encourage Denver Water and West Slope water entities to pursue mediation instead of long and costly court battles. Under the historic accord, the West Slope agreed not to oppose Denver Water's expansion of Gross Reservoir. In return for Denver Water diverting more flow from the Fraser, the West Slope was promised environmental mitigation and enhancement for its embattled rivers and streams.

Before the ink had dried on the deal, Denver Water said it was satisfied with the amount of water it would be able to divert. The West Slope said it was pleased with the mitigation and enhancement measures Denver Water promised to pay for, such as reengineering a narrower Fraser River so its depleted flows could run deep and cool to keep trout alive. Neither side demanded a dramatic showdown. Neither side lawyered up. Instead, they sat down and talked out their differences. In Colorado, said to have more water attorneys than any other state, this compromise seemed as surprising as the Hatfields and McCoys playing horseshoes together at a picnic.

Age-old enemies shaking hands across the Great Divide made for a fine story. But behind the roar of celebratory cheers could be heard some grumbling. Skeptics questioned whether the Colorado River headwaters could be saved by the pact. Peace is a worthy goal, but at what price should it be achieved?

"If anyone is a loser in this agreement, it is the river itself," wrote Kendrick Neubecker in an opinion piece in *The Denver Post*. He pointed out the agreement didn't address the damage caused by more than 60 percent of the Fraser and upper Colorado rivers already being diverted to the Front Range. The agreement would allow Denver Water to take another 15 percent or more of the distressed river's volume. Neubecker wrote, "With that much of the native flows removed, making about 1 percent available for 'environmental enhancement' as this agreement does, won't go far to help the river, much less improve it."

He had a point. Claiming the river would be healed while three-quarters of its water was stripped away seemed like praising the clothes of a naked emperor. And, as Neubecker pointed out, diversions might not end with 75 percent of the river's native flow siphoned to the Front Range. He wrote, "Future diversions by Denver Water and others are not ruled out. Even with cooperation, the upper Colorado and Fraser rivers could still be drained of their last drop."

Does the Fraser have anything left to give? Similar to how the food and medicine that maintain a person's health become meaningless if that person is deprived of water, mitigation and enhancement efforts can't prevent a river from dying if it runs dry.

The Fraser's flow is already so diminished it cannot flush itself clean of the road traction sand that washes into its clogged bed. Those of us who navigate our cars and trucks along the icy hairpin turns of Berthoud Pass in winter are grateful for the sand that stops us from skidding off a cliff. But every car that crosses the Continental Divide at Berthoud Pass adds to the water quality problems in the Fraser. The Colorado Department of Transportation (CDOT) collects sand from roads when the snow melts, but a beach's worth ends up in the Fraser after washing downhill—some six thousand tons each year. Grain by grain it builds, mixing with natural silt to plug the river's current and smother its cobble bed. Deprived of spring flushing flows by Denver Water's diversions, the Fraser fails to purge itself. And what was once a clear, cold flow crowded with life turns stagnant, warm, and sterile.

Who is responsible for this destruction?

I HAD HEARD DAVE Little's name hissed by angry environmentalists so many times, when I went to meet Denver Water's director of planning I was half-expecting a madman who tortures kittens. Instead I met a thoughtful person who excels at the difficult work of ensuring a quarter of the state's population have a clean, reliable water supply—a person who believes diverting more water from the Fraser to the Front Range is the right thing to do.

Dave's job entails the mind-boggling complexity of planning fifty years into a future ultimately unknowable, so he plans for five separate futures based on careful analysis of demographic projections, climate change scenarios, extreme droughts, forest fires in critical watersheds, and a zillion other variables no one considers when they open a faucet and clean water pours out. It is true Dave would rather see West Slope water growing lawns on the Front Range than trout in the Fraser. But as far as I know, he has never tortured kittens.

Dave says, "When each side demonizes the other, finding common ground is impossible." In my conversations with Dave I have heard his frustration over people opposing projects he planned, but I have never heard him dehumanize his opponents. He says he respects their passion and dedication. When you get Dave started on the topic of the earth's finite resources, he sounds, oddly, more like an environmental activist than a resource developer. He insists he's been concerned about overpopulation ever since he read *The Population Bomb* in the 1970s. He is as strident in his critique of global over-population as any critic of Denver Water I've met. Dave's job is to provide water for the people in Denver Water's service area; he insists neither he nor anyone else in Denver Water has the power to prevent sprawl in places like Douglas County. He points to infill and density in the recent development of Denver as major factors in reducing water use in the city, on par in its per capita water consumption with Boulder—a city my fellow Boulderites will tell you is the most enlightened place on the planet.

At a restaurant in Boulder I once listened to an archenemy of Denver Water explain why the organization is evil as he devoured a giant cheeseburger. Beef is a true water whore: Many hundreds of gallons of water—thousands of gallons by some calculations—were used to produce the beef patty this fierce critic of Denver Water scarfed after giving public testimony on the dangers of expanding Gross Reservoir, while Dave Little sat silently in the audience taking notes.

Dave has been with Denver Water for thirty-three years, nearly one-third of the organization's entire life. Contained within his brain are reams of historical and technical information—and knowledge about how to negotiate. He says the engineering side of his job is easy compared to the people part. Diverting, storing, and treating water presents solvable problems. People, on the other hand, can seem like unsolvable puzzles.

Kirk Klancke told me that many years ago he walked into Denver Water's headquarters and Dave Little said to him, "What are you doing here? Everyone hates you." A few years later, when Dave heard Kirk's wife had died, he came up to Kirk and gave him a hug.

Dave and Kirk both have imposing physical presences and exude warmth and likability. And both have been battling over water for more than thirty years—as if they were doppelgängers of each other on opposite sides of the Great Divide.

"Dave Little is a compassionate man," Kirk said to me, and he made sure I wrote that down. Kirk understands villains don't inhabit the Denver Water offices on the other side of the Divide. He realizes they love bringing water to the places they live as much as he loves keeping water in his community. There have been fierce battles, to be sure. Both sides have fired threats and accusations back and forth. Kirk, like a soldier drawn into a war not of his making, has fought for decades for what he believes is right. And when the smoke clears, he has moments of pause when he realizes people on the other side are doing precisely the same.

~~~~~~~~~~

WHILE GIVING ME A tour of the Fraser Valley, Kirk takes me to the Moffat Tunnel. The stone walls of Kirk's house hold pieces of the continent's Precambrian core, blasted loose when this historic train tunnel and aqueduct broke through mountain walls.

In 1921, after a flood rampaged through the city of Pueblo, killing 1,500 people, Denver's struggle to finance a tunnel through the Continental Divide ended. Denver released funds to provide emergency relief to its flood-ravaged rival; in exchange, Pueblo supported legislation authorizing the issuance of bonds for the Moffat Tunnel.

I've never seen this view of the tunnel Kirk now shows me. We stand in front of the water pipe, painted the green of mint chocolate chip ice cream. From this perspective, the water pipe dwarfs the train tunnel—which is fitting because Kirk is a self-described "pioneer bore skeptic."

A pioneer bore was drilled alongside the main train tunnel bore, presumably to guide the drilling crews and shelter them in the crosscuts between the two tunnels during blasting. But after the bond issue paid for the railroad tunnel, and then federal money turned the pioneer bore into an aqueduct, Denver got "the cheapest water diversion in the country," according to Kirk. He believes the water pipe was part of the master plan of David Moffat, one-time owner of Colorado's largest private water company—the precursor of Denver Water. Kirk jokes,

"Isn't it an amazing coincidence that none of the other train tunnels in Colorado have a pioneer bore?"

Lending credence to Kirk's theory, author Edward Bollinger in his book *Rails That Climb* interviewed Art Weston, an engineer who wrote an article in 1922 titled "Why Two Tunnels." Weston states, "Of course I knew that this 'pioneer' tunnel was intended to be used to bring water from the West Slope to Denver for domestic purposes, but they did not come out in the open and say so."

Distrust of Denver Water in Grand County is as deep as the tunnel that passes through the dark heart of the Great Divide. But putting aside suspicions of dishonesty on the part of Denver Water, historians agree that federal Public Works Administration money during the New Deal era was instrumental in turning the pioneer bore of the Moffat Tunnel into an aqueduct.[27] This detail may seem trivial, a crumb at the banquet table of history. Far from it.

Denver Water spends considerable time and effort defending its water rights in the Fraser River watershed. But the organization didn't "perfect" these water rights on its own. (Water rights are "conditional" until water is diverted; not until a diversion is made are the rights "perfected" and made absolute.) Denver was nearly bankrupt in 1933. Only with federal money could the destitute city transform the Moffat Tunnel pioneer bore into an aqueduct, thereby diverting water and perfecting the water rights that have allowed the organization to drain the Fraser Valley. This situation came about through a combination of state water law, federal water welfare, and another of Kirk's "amazing coincidences."

The Public Works Administration regional director happened to be George Bull, an engineer once employed by Denver in its quest to wrangle West Slope water across the Continental Divide. Mr. Bull made sure the federal government provided the money Denver needed to create the water tunnel, allowing

27   See *Denver: Mining Camp to Metropolis* by Stephen Leonard and Thomas Noel and *A Ditch in Time: The City, the West, and Water* by Patricia Limerick.

the city to swell into a major metropolis while it sucked the life from the Fraser watershed.

As with the Colorado-Big Thompson Project, American taxpayers footed the bill for infrastructure that redistributed water and wealth. Perhaps the Moffat Water Tunnel, a structure all Americans paid for, should be managed to reflect the interests of all Americans. Would preserving the health of the Fraser River Valley, a place of legendary beauty that inspired the leader of the free world to make it his Western White House, benefit the nation more than turf grass in Denver?

THE LAST PLACE KIRK TAKES me on our tour of the Fraser Valley is a park that features a giant bronze sculpture of President Eisenhower, his fly rod bending in an arc as he scoops a streamlined trout into his net. Kirk raised the funds to create this larger-than-life-size work of art to remind people the Fraser is a national treasure.

At the sculpture's base is a plaque. The first few paragraphs explain the history of President Eisenhower fishing and performing the duties of his office in the Fraser Valley. The last lines were written by Kirk: "The Fraser River you see today is not the same river that attracted Ike in the 1950s. As Colorado's Front Range communities grow, more of this river is diverted to accommodate growth. We must find a balance between the needs of Colorado's growth and the needs of Colorado's natural environment…which attracted Ike then and millions of visitors each year today."

Around the statue trickles a stream, weed clumps swaying like medusas in the weak current. As I ask Kirk a question, I realize he has moved out of earshot and is lying in the grass. He stuffs his arm up to his elbow into a clogged pipe to scoop out gobs of mud and debris. When he stands back up, he laments the lack of maintenance around this memorial. He wants young people to see the sculpture so they understand what's at stake in the Valley of the Fraser. The unclogged pipe chugs water into the stagnant flow around the statue, stirring up sediment, blasting apart tangled weeds. After a few minutes of muddy turmoil, the stream runs strong and clear: a powerful flushing flow, like the Fraser needs.

The Fraser will never again function as it did before Denver Water began diversions. But with clever engineering and sound management, the harmed river could be restored to some approximation of health. To give me a glimpse of what a revitalized Fraser River would look like, Kirk shows me a section reconfigured by a habitat improvement project completed in 2005.[28] The project narrowed the river, forcing it to run faster and form deeper pools for fish. Biologists explain the reconstructed channel is functioning well in transporting sediment, and the fish are healthy because the bugs are thriving. As Kirk and I walk along the bank, trout send rings spreading across a deep pool, and the earth beneath our feet crackles with the paper-thin skins left by stoneflies as they transformed.

This redesigned river reach offers proof of concept. If all goes according to plan, the ailing river will be turned into a healthy creek. Whether this diminished waterway can maintain a functioning ecosystem along its entire length and into the future is an open question.

BUD HAS KNOWN OF Kirk's work on the Fraser for several years. These allies marshal forces in common cause, for the Fraser's fate is closely bound to the Colorado. If the Fraser dies, like a gangrenous limb it will spread its infection into the upper Colorado River.

Kirk serves on the board of Bud's Upper Colorado River Alliance, and they share an abiding respect for a man Kirk calls "a fallen comrade": George Beardsley. As mentioned, George had joined with Bud to create the Upper Colorado River Alliance. Besides witnessing the devastating changes to the Colo-

---

28   The town of Fraser partnered with Trout Unlimited and other entities to create the Fraser River Enhancement Project (FREP).

rado River headwaters, George possessed insider knowledge of water politics from serving on the Denver Water Board from 2004 to 2009.

George was known as an independent thinker who wouldn't rubber-stamp whatever Denver Water's planning department put in front of him. He criticized the scheme to expand Gross Reservoir and fill it with more water from the Fraser. Bud says, "George didn't think the project was really needed. And he was worried about the damage it would do to the Colorado." Kirk and Bud honor George's memory by continuing his tradition of skeptical appraisal of dubious claims made by Front Range water providers.

Part V

Chapter 18

LEARNING BY DOING

I n March 2014, a much-heralded agreement—the Mitigation and Enhancement Coordination Plan—marked another truce in the battle over the Colorado River headwaters. Grand County and Trout Unlimited agreed not to oppose Denver Water's Moffat Project, which would divert more flow from the Fraser River watershed. And Denver Water agreed to implement a program to help repair the damaged river.

The linchpin of the agreement is "Learning by Doing." This adaptive management strategy calls for regular monitoring of the river to assess its health by measuring stream temperature, studying riparian vegetation, and counting aquatic macroinvertebrates (little critters with no backbones that live at least part of their lives in the water and can be seen with the naked eye—stoneflies, for example). According to the agreement, if environmental problems are detected, Denver Water will provide specified amounts of money and water to improve the river's health. And Denver Water will divert additional flow from the Fraser only during periods of high water.

Learning by Doing grew from the realization that changes to rivers cannot be predicted with precision. Windy

Gap Reservoir devastated the Colorado's health by armoring the riverbed downstream of the dam—yet this effect didn't elicit major concern during the years of environmental debate that stalled the project. If Denver Water and Northern divert even more flow from the Colorado River headwaters, the ecosystem will be shoved closer to a tipping point, but we cannot anticipate the precise repercussions. No precedent exists for how to maintain the health of a river with three-quarters of its water removed.

Chinese leaders characterized their nation's economic reform, which progressed without a blueprint, as "crossing the river by feeling for the stones." Learning by Doing is water policy reform without a rigid blueprint for how to proceed. It is a commitment to move forward by feeling for the stones while crossing to the other side—from ecological ruin to river restoration. Critics complain ambiguity weakens the agreement; proponents insist regulatory flexibility will prove essential when facing unanticipated challenges in the coming decades—they explain that Learning by Doing is ideally suited to an era of climate change.

Using adaptive management to try to bring a river so severely damaged back to health is an experiment of great consequence. If Learning by Doing saves the Fraser, the strategy could have far-reaching implications for other severely impaired rivers around the nation, and perhaps the globe. If, however, the strategy fails, the Fraser will continue to trickle toward death, and the state of Colorado will be forced to explain to the country, and to the world, how it killed a river cherished by one of its nation's presidents.

Key to the agreement is Denver Water's multipart diversion system in the Fraser River Basin, which allows flexibility in moving and managing water. As Bud says, "Denver Water has the ability to capture water ten ways from Sunday." Denver Water will use this flexibility to take pressure off stretches of stream that scientists decide are struggling. The organization will shuffle the management of its diversion systems to make streamflow adjustments that keep temperatures within a safe range for aquatic life and provide flushing flows that maintain a healthy environment. Denver Water will also devote dollars for stream improvement efforts similar to the channel-narrowing reconstruction project on the Fraser that Kirk showed me.

But how much water can a river afford to give? Is the 75 percent of native flows that will be diverted too much? Few businesses would survive if three-quarters of their annual revenue were stripped away. An experiment that removed 75 percent of the water from a human body would not end well.

WHEN THE MITIGATION and Enhancement Coordination Plan was announced, all parties involved slapped each other's backs so hard they practically left bruises. Jim Lochhead, Denver Water's CEO and manager, declared, "This plan represents a new, collaborative way of doing business together when dealing with complex water issues." Trout Unlimited called the agreement "a victory for the river." Lurline Curran, Grand County's manager, commented, "To all parties' credit, this effort has succeeded." Kirk Klancke stated, "I'm gratified that this agreement keeps our home waters healthy and flowing." At a meeting in Grand County when the agreement was announced, tears streaked the cheeks of this stonemason who for three decades has served as a steadfast voice for the Fraser.

Learning by Doing is also the linchpin of a mitigation and enhancement agreement with Northern Water. Skeptics point out Denver Water and Northern have had twenty years of learning by doing with Windy Gap and the Moffat System—but they don't seem to have learned much about maintaining the health of a river.

Bud says, "Everyone thinks the oil companies are bad, but they're no worse than the water buffalos." He wonders if the Learning by Doing agreements with Denver Water and Northern rely too much on the good faith of organizations that have in the past done everything they can to avoid taking responsibility for sound stewardship of the river. Barry Neh-ring worries that only when Northern and Denver Water have their backs to the wall will they open their checkbooks. He wonders why they quibble over paying to repair the damage they've done. He questions why these organizations with the power to make things right on the river don't just do a one-time assessment on their water customers to pay for mitigation to fix what they broke in the upper Colorado watershed. He has an interesting point. Many people fighting to save rivers have their own version of this plan.

Kirk Klancke, for example, points out the cost of reviving the Fraser cited in one report—$7 million—could be paid for by charging Denver Water customers 54 cents a year for ten years. Kirk proposed a voluntary program similar to one ener-gy utilities use: Customers can round up their bill to the near-est dollar, and that spare change adds up to a substantial fund used to finance a good cause. What's not to like about that?

"It will never happen," Kirk was told by a Denver Water em-ployee.

Denver Water and Northern tend to shoot ideas like these down as though they were clay pigeons. But why water provid-ers didn't adopt Kirk's plan a long time ago instead of spend-ing a small fortune on public relations messaging is unclear. Perhaps they could have fixed the river for less than they've spent trying to convince everyone they intend to fix the river.

Denver Water serves more than 1.3 million people, each of whom has participated in damaging the Colorado River head-waters. If every customer in the Denver Water service area was assessed a one-time "Colorado River Restoration Fee" of, say, $10 (or $1 a year for ten years), $13 million would be raised. That doesn't include Northern with its vast service area. For the cost of a couple of lattes per person spread over a decade, water providers that harmed the river could have healed it years ago instead of trifling over dollars to finally do what des-perately needs to be done.

Bud says, "I learned a long time ago in the oil and gas business that it's better to do the right thing for the environment up front—or else you end up paying a lot more down the road."

Adequate mitigation paid for by water providers who divert from the river should be a cost of doing business. It shouldn't be a debate. It's ludicrous that we even have to discuss it, let alone fight over it and write books about it. Bud, Barry, Tony, and Kirk should be on healthy rivers fishing, and I should be in Brazil writing about that nation's meteoric rise. Yet here we are, fighting this strange battle in our backyards, squabbling over adding a dollar per year to the water bills of people who receive cheap water in a wealthy state in the richest nation in the history of the world.

The cost of a cup of coffee, the price of a few lattes: Pick your metric, but the central truth remains the same. All this strife could have ended for what amounts to pocket change. While the Colorado River dies, we send our best and brightest across the globe to teach other nations to take care of their resources. And they come here to learn from us.

Jim Lamont, executive director of Vail Homeowners Association, is as fierce a critic of Colorado's water policies as anyone I've met. He told me he once saw a man wandering at the end of his road—not unusual on most roads, but Jim lives on a mountaintop far from the madding crowds. He says, "I asked the man if I could help him. I thought he might be up to no good. Turns out he was a Buddhist." He had traveled from Tibet to learn how we manage our natural resources— he wanted to apply the lessons he learned here to managing the liquid treasure that spills from the snows of the Tibetan Plateau, nourishing ecosystems and civilization across a vast swath of Asia.

It seems a sort of parable. A man from Tibet makes a pilgrimage to the headwaters of the Colorado, our Mother of Rivers, seeking wisdom at the source. Jim Lamont, as big a curmudgeon as I've met in the mountains, didn't have the heart to tell this poor pilgrim his journey had been in vain. We have a long way to go before our stewardship of rivers is worthy of study—other than as a cautionary tale of how to waste what is precious and rare.

But the state of Colorado now has the opportunity to make up for its past mistakes mismanaging one of its most prized natural assets. It can lead the New West in a forward-thinking water policy to protect the headwaters of the state's namesake river.

I HAVE HEARD WATER providers refer dismissively to Bud and others trying to save the Colorado River headwaters as "a few old white guys who want to catch trout." Mely Whiting, staff attorney for Trout Unlimited, laughs when I tell her this and says she is "a middle-aged Hispanic chick." She also insists she's "just a beginner" at fly-fishing. For eight years she has been working with TU—not because she wants to make sure rivers are filled with trout for rich white guys to catch but because she cares about conserving natural resources. She understands that when you save creatures finely tuned to their aquatic environment, you save rivers.

Mely's pet peeve is when activists show up to protest a project and then disappear when the battle is over. Her point is an important one. To complain about a dam takes little imagination or courage. Screaming *no* at the top of one's lungs is seldom productive in a world where water and energy issues become increasingly complicated as more and more people fill the planet. To be *against* something is easy. To be *for* something is hard. Mely understood early on she had to sit down with the opposition and listen to their needs. Then she had to find ways to make some concessions while maintaining a hard-nosed negotiating style that protected the needs of the river. To forge an agreement that would bring several parties together in partnership to restore the watershed, she had to be literate in law, requiring years of education in formal classrooms and in the real world.

Mely cut her teeth on water law by working for the firm that represented Denver Water during the Two Forks controversy. After a career that spanned working for private, state, and federal entities, she sent her daughter off to college and then teamed up with a nonprofit so she could apply her legal acumen to protecting sentinel species—which she is less interested in catching than conserving.

Kirk told me Mely checks the readings of river gauges on the Internet and calls him when anything looks amiss. Like Bud and Kirk and so many others, Mely has made the river's health her personal charge. She is as fierce a protector of rivers as Glenn Saunders was a staunch defender of Denver's interests on the West Slope. And Mely is a driving force behind what may someday be considered one of the most impactful agreements in the history of Colorado water.

PHOTOS OF HISTORIC WATER deals in Colorado show nothing but white guys—rooms full of Caucasian men signing papers that control the power of water. In contrast, two of the champions of the Mitigation and Enhancement Coordination Plan, along with Mely Whiting, are Barbara Green, Grand County's environmental counsel, and Lurline Curran, Grand County's eminently quotable manager.

"Don't goddamn come here anymore," Lurline once said, summing up the West Slope's position toward the Front Range's search for new water sources. Spoken at a Colorado River Basin Roundtable meeting in 2013, these words went viral in the Colorado water world. Lurline, a white-haired grandmother, told me, "I have a minor in art, and cussing to me is an art form."

Lurline met with me before she headed to a meeting with the Bureau of Reclamation to negotiate a carriage contract on Grand County's behalf. If you don't understand carriage contracts, well, you're not alone. Only a handful of people can fathom how Windy Gap Firming water relates to carriage contracts. Lurline is one of them—and she has no formal training in water law. Carriage contracts are the kind of wonky stuff that water attorneys debate in a language you would swear is as foreign as Farsi.

When Lurline isn't swearing, she is speaking the language of western water. To get a flavor for this strange tongue, but to avoid putting you into a stupor, here's an excerpt from a Grand County Planning Commission meeting:

> The 2012 WGFP proposes a different operational regime for the Windy Gap water rights known as prepositioning. Prepositioning was never contemplated or permitted with the original Windy Gap Project and it changes the timing and magnitude of diversions from those originally taken into account by Grand County.... Prepositioning requires an amendment to the existing Amendatory Contract (Carriage Contract).

Lurline is undaunted by this kind of tediously complicated technical detail. She is self-effacing to a fault: She credits her commissioners and the Colorado River District with giving her the latitude to negotiate agreements—but Mely and others insist Lurline has been a powerful force in crossing the divide between West Slope communities that rely on healthy rivers and Front Range water providers that depend on diverting those rivers.

Studies show that in general women are better than men at collaboration; women are also more likely than men to give credit to others rather than pointing to themselves as heroes. If Colorado is moving toward a new water era by replacing conflict across the Great Divide with collaboration, it makes sense that women are helping lead the way.

Lurline having a psychology degree also makes sense. Scarcity of a resource we cannot live without has a way of distilling human nature to its most basic elements: warring and peacemaking, competition and cooperation. Lurline says that during nearly a decade of negotiations with Denver Water, she formed unlikely connections with adversaries across the table. Her arguments with Dave Little are legendary—the two scrapped like wildcats. But Lurline says, "Dave and the rest of the Denver Water crew weren't bad people. They just had a job to do." Lurline falls silent a moment. "People became people to us. We learned to see the goodness in each other."

Lurline went through twenty weeks of chemo for breast cancer during the negotiations but never missed a meeting. Even when she had four surgeries in four weeks, she showed up at every negotiation. She says, "As long as I kept doing my job and I didn't look in the mirror, I felt okay."

The negotiations helped Lurline keep her mind off her sickness by giving her a goal to focus on. She says, "I wanted to leave something behind for the county where I'd been raised and where my grandchildren live." Her father had owned a machine shop in Kremmling; as a girl she had played in local rivers. "I never dreamed they could dry up," she says, her white hair bouncing as she shakes her head.

Lurline explains she has picked out an assisted-living facility in Grand County, near where she now resides, and not far from where she spent her childhood. "When I'm crazy as a loon and rocking on the porch of the assisted-living center, I want my grandkids to say I did something important."

The Mitigation and Enhancement Coordination Plan that Lurline negotiated with Denver Water is not perfect, and neither is this grandmother with a habit of hotheaded cussing—Lurline freely admits both points. But it is impossible to spend time with Lurline listening to her tell the story of the agreement and not believe she did what she thought was right for the place she loves. She tells me, "My one wish is that future generations in Grand County will continue to pay attention to preserving the rivers." In this way Lurline is like Bud, and even Mike King.

The likelihood of Mike King and Bud casting to trout together on a river is laughably low. But Lurline says of Dave Little, her long-time nemesis, "He's a good man." Though their beliefs about the best use of West Slope rivers are as different as fish and faucets, the bond they formed allowed them to work toward replacing a past of bitter conflict with a future of cooperation.

If Lurline's wish comes true and the Colorado River head-waters are healed through the cooperation of Denver Water and Grand County, and Lurline's grandkids continue to keep the river flowing and healthy, the agreement Lurline and Mely forged will form an important new chapter in the story of Colorado water. In contrast to the photos of the men who wrote the 1937 legislation that paved the way for the engineering triumph of the Colorado-Big Thompson Project, photos of women will illustrate this new chapter of sharing water.

Lurline grins and tells me, "You couldn't make this shit up."

THE FINAL ENVIRONMENTAL IMPACT statement (EIS) for the Moffat Project is astoundingly huge: including all its appendices, it comprises some 16,000 pages. That's not a typo. That's sixteen *thousand* pages.

Similar to how the human brain is baffled by the vast span of geologic time, our minds shut down when faced with the staggering length of this study. The table of contents is sixty-two pages long. The executive summary is an eighty-page beast. If all the documents of the final EIS were stacked atop each other, they would tower six feet tall. Organizations that requested printed copies were told by the Corps of Engineers it was too large to print. Jokes were made about an EIS being done on the amount of paper needed to print the EIS.

It would take a person many months, if not years, to read through the 16,000 pages and make sense of the material. The period for public comment after release of the final EIS was forty-five days—and the Corps refused to extend the period, even when the Boulder County Board of Commissioners, which must assess the impacts of the largest construction project in the county's history, asked for more time. Even Denver Water asked the Corps for an extension so everyone could try to make sense of the behemoth study.

In a Boulder County Commissioners hearing on Gross Reservoir expansion after the final EIS was released, one member of the public said he'd calculated that reading the whole document in the forty-five-day timeframe allotted by the Corps would be like reading a 355-page novel—a really boring novel—every day for forty-five days, leading him to suspect the public comment period was a ruse by the Corps. During the three hours of testimony at the meeting, citizens delivered harsh criticism of the project, ranging from a disgruntled former employee of Denver Water calling the organization "a corrupted public utility run by its developer friends," to a professional truck driver who testified to the danger of navigating construction trucks up the narrow, winding mountain road to the dam-building site. A scientist punctuated the dry facts of his presentation with an eye-popping picture of a Rocky Mountain iris, a seasonal wetland plant that relies on high river flows for survival. He explained that along riparian corridors a million of these flowers might be blooming right now—and every one of them is at risk if their habitat is degraded by further depletion of the upper Colorado River watershed.

One theme that came through loud and clear in the three hours of testimony to the Boulder County Commissioners was this: Saving a watershed is about much more than protecting trout to catch. It is about saving countless creatures, including ourselves—and science reveals these connections. When we prevent rivers from swelling with spring snowmelt, the consequences ripple outward from their shores. Trout are a proxy for entire systems upon which the wealth of wild nature depends.

IN PRINCIPLE, THE 2014 Grand County Mitigation and Enhancement Coordination Plan points to a healthier future for the upper Colorado River watershed. But in practice, the agreement is filled with statements like this: "Denver Water will use reasonable efforts to provide water on an as-available basis to help achieve the desired flows."

Bud scoffs at the word "reasonable." As an oil and gas guy, he knows all about contracts—what makes them legally binding, what makes them not worth the paper they're printed on. He explains that "force majeure" is standard legalese that allows a party to suspend or terminate its obligations when circumstances beyond their control arise—a drought, for example. Or perhaps a doubling of population on the Front Range. A cynic could imagine Denver Water's attorneys rejoicing at the use of "reasonable," for shrewd legal minds can stretch this elastic word into whatever shape they choose.

As the metropolitan area continues its snowballing growth, what will be considered "reasonable efforts to provide water on an as-available basis" to support the river's health? As mentioned, Senate Document 80, approved by Congress in 1937, stated that the Colorado-Big Thompson Project must be operated "[t]o preserve the fishing and recreational facilities and the scenic attractions of Grand Lake, the Colorado River, and the Rocky Mountain National Park." This provision has been blatantly ignored for Grand Lake and the Colorado River. Furthermore, the Colorado-Big Thompson Project was authorized by Congress to provide supplemental irrigation water, but the project has been used to support urban growth and suburban sprawl. When it comes to enforcing regulations on the upper Colorado River, Bud says, "The sheriff is asleep. Why do we pass laws if we're not going to enforce them?"

David Nickum, executive director of Colorado Trout Unlimited, says he likes the realpolitik principle of "trust but verify" popularized by Ronald Reagan during the Cold War. He explains that TU is working to get the mitigation and enhancement plan tied to the permit Denver Water needs from the Army Corps of Engineers. If down the road an arbiter decides Denver Water isn't complying with Learning by Doing, the Corps could suspend Denver Water's permit. Dave explains, "This doesn't only provide a fail-safe if worse comes to worst. The unpleasant prospect of having to reopen the mitigation question gives Denver a real incentive to work hard to ensure

Learning by Doing's success. All the parties at the table have a strong incentive to put in their best efforts—which makes it easier to be hopeful about what this agreement can achieve."[29]

~~~~~~~

KIRK KLANCKE AND MELY WHITING, currently serving on the Learning by Doing committee, are making sure baseline data is gathered for the Colorado headwaters. Establishing river conditions before Denver Water begins its Moffat Project allows future impacts to be measured—giving vigilant watchers a tool they can use to guard the health of the headwaters.

Like all agreements that protect a natural resource, Learning by Doing will only work with constant vigilance. The river needs its next generation of defenders to make sure hard-won protections are carried into the future.

Dr. Gene Reetz, a retired EPA employee who played a crucial role in the veto of Two Forks, says he nearly fell out of his

---

29   Dave, whose quiet intelligence and diligent defense of rivers commands attention, further reflects on Learning by Doing: "The collaborative spirit is commendable and hopefully transportable to other situations, but I would never suggest that environmentalists accept 75 percent depletions in a river in exchange for adaptive management on a system that wasn't already grossly over-depleted and in major ecological decline." On this point he is adamant. "The agreement for the Colorado only makes sense in its context. The Fraser and Colorado were dying river systems, and TU came to the conclusion that saving them required a new approach instead of just saying no to more diversions—one that brought Denver Water and Northern to the table as partners with their significant resources and system operations as part of the team working to conserve the river. And that would only happen if they had something to gain from the process as well. Much of the negotiation efforts that Lurline, Barb, and Mely navigated so skillfully were trying to create the right incentives for everyone, so that all the key parties had something to gain from the Colorado River Cooperative Agreement and Mitigation and Enhancement Coordination Plan—and they all had something to lose should it fail."

chair when he attended a presentation by Denver Water and a spokesperson for the organization mentioned Two Forks Dam isn't necessarily off the table.

In the fight over Two Forks, the clear winners were TU and other environmental organizations; Denver Water was the definite loser. However, the historic veto stopped the project as proposed; a new version of the dam could be revisited in coming decades. Under the South Platte Protection Plan adopted in 2004, Denver Water agreed to a twenty-year moratorium on development of the Two Forks site. When the moratorium expires, a project will be possible. Simply put, Two Forks Dam is not completely dead, nor will it ever be.

Gene says, "I can't help but reflect on the seemingly inherent injustice that environmental victories are at best temporary whereas defeats are permanent."

Perhaps Learning by Doing represents a new kind of environmental victory—one without winners and losers. A victory that allows former adversaries to work together toward the goal of healing the river. A victory that channels the extraordinary effort required to compile thousands of pages that document a river's decline into a proactive program to save it.

Chapter 19

THE LONG HAUL

One afternoon, while Bud and I talk at his home in Denver, I ask him how he feels about all the flack he gets from environmentalists over fracking. He sighs and shakes his head. "A lot of greenies live in cities and don't really understand the natural world. When I was a boy, I wandered around in forests and fields with my .22. There were no environmentalists—I didn't even know what those were. I just knew about the animals and the landscape and the weather."

Bud falls silent a moment. "Basically environmentalists are selfish people—they are trying to preserve the world they're living in at the time they're living in it. I'm an earth scientist. We study geology that goes back millions of years. This helps us interpret modern data and consider where to drill. It's a very risky business. Our interpretations can be wrong. We have to figure out the mechanics of drilling to the target. And things can go wrong when testing and completing a well. We get labeled as *exploiters*. We prefer the term *explorers*. Everyone uses what we as exploiters have provided. We only provide what people demand."

Bud pauses a moment, then continues: "It's like the water buffalos. They bring water from the West Slope to the Front Range because people here want it." We're sitting in Bud's kitchen, which has a view of the green grass in his backyard. Along with his sprawling lawn, there are other tensions I cannot ignore. On the upper Colorado River, state law allows landowners to kill beaver and blast apart their dams. This busy critter is integral to the ecosystem, but its waterworks can screw up the fun of fisherman. And anglers, including me, enjoy casting flies to species of trout not native to the watershed. The cutthroat trout stocked in the river by nature's design don't fight as spectacularly at the end of a line as brown trout from Eurasia or rainbow trout native to the Pacific Rim.

But there is no going back: The river's natural systems have been so thoroughly redesigned by dams and diversions and nonnative fish, reintroducing the cutthroat trout species to the upper Colorado wouldn't work. The river will never again be what it once was. Its best hope of being restored to some semblance of health is Bud's battle to make sure rainbow trout once again thrive in the river. This is an imperfect battle on an imperfect river in an imperfect world. But it is a battle worth fighting. With river management, as with energy development, making perfect the enemy of good gets us nowhere.

"I love environmentalists," Bud tells me. "They keep the pendulum from swinging too far in one direction. And I love the developers because now I have an iPhone and can be in touch with people, and I can hop in a plane and go see things around the world that people have dreamed of. So I'm selfish, too. I want the river to be healthy for my grandkids. I want them to be able to watch a dipper bird dive under the water and come up with a big stonefly in its beak."

Bud looks his seventy-two years today. He has recovered from broken ribs after being bucked from a horse, but his wife just had a total knee replacement, and he has been helping her—"playing nurse," he says. Lately I've noticed holes in Bud's memory for conversations we've had. And I have been facing my own deteriorating ability to recall details. My knees that once carried me over mountains with ease now shoot bolts of pain when I stand. Sometimes when I feel that pain, I understand the time I have left on this earth will probably be shorter than the span I've already spent. The great problem of being human is, of course, foreknowledge of our own death. Saving a river offers us a bit of immortality. While our bodies and minds erode, the river rolls on. Knowing a river's flowing waters will be restored to health helps bridge painful gaps in memory and soothes our aching joints. Knowing the upper Colorado will again fill with stoneflies and trout, as it did before our farms and cities sucked the river dry, and knowing this abundance will one day be witnessed by children not yet born, offers some solace to minds being whittled down by time.

Bud says, "When I was recovering from rotator cuff surgery last winter, every day I went on snowshoes with one ski pole to a river. I'd see how many fish were under the bridge. I watched a dipper bird diving into the water. A big bald eagle would come and perch in a tree and watch the river. He was going to harvest the fish. Just like the little dipper bird was going to harvest the bugs in the river. Just like I'm going to keep harvesting oil and gas so I can do the things I enjoy and provide something for other people so they can do the things they enjoy. And I'm going to keep fighting for the river so that it's healthy when I'm gone."

Bud taps a finger against his kitchen table. Motes of dust dance in slanting light. He says, "We want the world to stay a certain way forever. We want to fix things in time and place, and we don't want them to ever change. We just want to look at our lifespan and our children's lifespan. But for mother earth, the lifespan is measured in billions of years. The earth is always changing. This is a living planet. The forces of nature far exceed anything we can do. And we're so young. We're not even the dust on this table in the lifespan of the world."

Bud was working in the Berthoud oilfield in Colorado just before the Big Thompson flood. On July 31, 1976, bloated thunderheads dumped a year's worth of rain—fourteen inches—in four hours. A wall of water twenty feet tall crashed through

Big Thompson Canyon, forming a flash flood that killed more than 140 people and destroyed more than 400 homes. A massive steel siphon, a component of the Colorado-Big Thompson Project, was ripped loose and carried away like a twig. A few weeks previous, Bud had cast flies into pools of the peaceful canyon.

All of us, the enviros and the oilmen, want to hold on to the world as it is. We want to preserve it as long as we can, until the inevitable cataclysm buries us in the mystery of this ever-changing Earth riddled with oil and water. Our lives in geologic time are as quick as the click of a camera shutter. What we have now we want to be fixed in time forever. We want the river to keep flowing, the dippers to keep diving, the eagles to watch from the trees. And because so many of us want to live in a place where we can experience these wonders, we have manipulated rivers on a staggering scale.

WHEN CONSTRUCTION OF THE Colorado-Big Thompson Project began in 1938, the area served by Northern Water contained some 75,000 people. When the project was completed in 1957, population had doubled to 150,000. Now it has soared to more than 860,000. And population increase along the northern Front Range, one of the fastest-growing areas in the nation, shows no signs of abating.

A consultant for Broomfield, the biggest participant in the Windy Gap Firming Project, stated in a *Denver Post* story, "We can't sell taps until we have a reliable supply. That water is a resource for future development."

How much development can the chronically dry Front Range sustain?

Opponents of the Windy Gap Firming Project hear the sound of yet another straw sucking the upper Colorado River dry. Supporters hear water flowing from all the new faucets on the Front Range. They point out that the gap between water supply and demand in Colorado is widening, and they insist that as the population continues to expand, this gap must be filled with more water diverted from the Colorado River into Chimney Hollow Reservoir. Ten cities, two rural water districts, and a power provider are counting on that diversion to help them meet their growing water and energy demands.

To get the reservoir built and more flow diverted from the upper Colorado across the Continental Divide, Northern once again has had to fight its way through a series of barriers, from navigating the environmental impact statement (EIS) process to negotiating a carriage contract with the Bureau of Reclamation. Northern also entered into negotiations with Grand County's Board of Commissioners to approve what's known as a "1041 permit."

Granted through House Bill 1041, "1041 powers" give counties in Colorado control over local land use. In a nutshell, the legislation provides counties with the authority to approve or deny development proposed by entities outside their boundaries.[30] But Northern isn't proposing to build anything new in Grand County with its Windy Gap Firming Project—it wants to create the Chimney Hollow Reservoir southwest of Loveland to store Windy Gap water on the East Slope. Brian Werner, public information officer for Northern, told me the folks at Northern could have rolled up their sleeves and refused to enter 1041 permit negotiations with Grand County. But Brian insists this kind of combative attitude isn't the Northern way anymore. He says, "Leaders at Northern realize they have to listen to the concerns of all stakeholders. They have to try to find collaborative solutions."

The EIS that Northern had to prepare for the Windy Gap Firming Project will cost the organization some $15 million—which gets Northern either a yes or a no from the federal government on moving forward with the project. Northern doesn't want disgruntled groups hiring attorneys to stop their Windy Gap Firming plan. So Northern talks with

---

30   1041 powers may play a critical role in Boulder County's decision whether to oppose Denver Water's controversial plan to expand Gross Reservoir.

stakeholders concerned about the project and offers them concessions. A cynic could say this is diplomacy not at the point
of a gun but at the tip of a pen in the hand of a water lawyer.

Whether by generosity or calculated self-interest, Northern
did willingly participate in the 1041 permitting process with
Grand County. Through extensive public hearings, Northern
addressed concerns about the Colorado River's health and the
deteriorating clarity of Grand Lake. Northern sat down at the
table with Grand County and Trout Unlimited and pledged
$250,000 to research a bypass to restore the river's health.
Northern also agreed that if the bypass was found to be beneficial, it would put another $2 million toward building and
maintaining it. Satisfied with these agreements, Grand County approved the 1041 permit, allowing Northern to pursue
its plans to divert more water from the river. Fighting to restore the Colorado's health is not easy, nor is it easy to provide
water to the farms, industries, businesses, and homes of the
ever-growing Front Range.

WHEN I MEET WITH Brian Werner at Northern headquarters in Berthoud, I mention the disappearance of the mottled
sculpin downstream of Windy Gap. He shakes his head. "You
know, when I was growing up in Colorado, we used to just kill
suckers. If we caught one, we'd bash its head on the rocks and
throw it on the bank. Now we're spending millions of dollars
to save them." He doesn't seem to know the difference between
a sculpin and a sucker—and he doesn't care to. Brian likes to
play golf. After thirty years at Northern, he's close to retirement. It's safe to say he will not be studying the mottled sculpin in his leisure time.

As public information officer, Brian's job is to tell Northern's story. He says the organization now recognizes it cannot
ignore the environmental consequences of delivering water to
its customers. "We can't just talk about the importance of water conservation," he says. "We have to walk the walk."

And so, Brian leads me on a walk around the conservation gardens of Northern. He points out a clever water feature: a map made of stones and ponds illustrating the Colorado-Big Thompson Project. This model, unlike the actual C-BT, uses recycled water. In the gardens grow some sixty varieties of turf grass that need varying amounts of moisture. Xeriscape displays showcase drought-tolerant plants. In the distance beyond the demonstration gardens, mountains rise above the plains. Longs Peak, monarch of Rocky Mountain National Park, reflects sunlight from its snowy slopes. Brian explains that between Longs Peak and Meeker Peak, a snowfield shaped like an hourglass will appear later in the summer. Irrigators once looked to this snowfield to know how much water would be available for their crops. Now they look to Northern and its diversion of West Slope water through the C-BT to provide a steady supply.

The winter of 2013–2014 has been generous. The depth of the snowpack in the Colorado Rockies now stands well above average. Enough water will be available for agriculture and cities—and for the upper Colorado River watershed. But no one knows what the coming years will bring. In times of ample moisture we can all get along. The droughts are what test us. Colorado's state constitution gives priority in times of water shortage to domestic use, followed by agriculture, and then manufacturing. The Colorado River's health is not covered in the constitution of the Centennial State, which entered the Union in 1876.

Surveillance systems monitor Northern's gardens and the surrounding grounds. Prominent signs let you know cameras are focused on you. These electronic eyes keep close watch as you gaze past yuccas at the snow-covered peaks. Each frozen crystal in the snowpack is owned by someone or something—a farmer, a coal-fired power plant, the Republic of Mexico. The moment a snow crystal turns to liquid, delivery to its lawful user begins, and the meltwater is monitored as closely as someone visiting Northern's headquarters.

Brian and I leave the gardens that demonstrate sustainable water use and walk back inside the office, where coffee is served in disposable Styrofoam cups. On the walls hang historic photos, including one of President Theodore Roosevelt. Brian is reading a book about the twenty-sixth president. He explains that Roosevelt was the first president to understand the power of the media and use it to his advantage—he invented the bully pulpit. Brian's enthusiasm for this topic makes sense, given he understood many years ago the need to craft Northern's story to fit the narrative of the times.

On Brian's bookshelf sits a copy of *Cadillac Desert*. While researching the book, Marc Reisner visited Northern and talked with Brian. Brian says when *Cadillac Desert* was published, many of his colleagues refused to read it because they knew it was critical of water development in the West. Brian not only read the book, he memorized the page numbers where Reisner makes brief but positive mention of the C-BT Project.

To Brian's credit, he understood early in his career the world was changing and Northern needed to change with it. He was trained as a historian, unlike many of his peers in the water industry who come from backgrounds in engineering and law. He seems to earnestly believe the story he tells of Northern taking responsibility for environmental stewardship. Examples to support this narrative abound, from Northern's water conservation programs, to the reduced per capita water use in its service area, to the organization's willingness to negotiate with Grand County and Trout Unlimited instead of digging in its heels and fighting requests for mitigation to offset the impacts of its diversions.

Bud, however, points out that while all this goes on, the Colorado continues to linger on life support. To someone who has fought for fifteen years to prevent the river's decline, demonstrating varieties of turf grass and holding up collaborative agreements as evidence that a new era has dawned are not important. What matter are the disappeared stoneflies and the dead trout.

When I ask Brian about the upper Colorado's deteriorating health, he tells me his pet peeve: "The enviros bitch about protecting the rivers but then expect other people to pay for it." Fair point. But Bud has been paying out of his pocket for many years to bring the river back to life. And Bud is not the kind of enviro Brian is referring to. Bud grumbles to me about Boulder being "too green."

As spokesperson for an organization that counts Boulder as one of its customers, Brian is careful not to malign the city. Boulder serves as a hotbed of environmentalism, yet part of its water supply comes from the West Slope through Northern's C-BT Project. Boulderites with "Save the Earth" bumper stickers on their hybrids contribute to draining the Colorado River when they water their organic gardens.

I'd be willing to bet all the tofu I eat in a year that when Brian is on the golf course with his buddies, the words he utters about Boulder enviros would not be printable in a newspaper. And I can't blame him. We enviros have made his life pure hell at times. Nothing is easier than living comfortably on the Front Range because of Northern's efforts to provide water—and then criticizing the organization for providing that water. Whether Brian literally pulled the hair from his balding head or it fell out because of people like me bashing the dams that allow me to live here, he is gracious enough—or at least diplomatic enough—to display my book *Dam Nation* on his desk when I visit. Had he used it for a doorstop, I would not have blamed him. Had its pages been substituted for toilet paper in the bathrooms of Northern, I would have understood.

Criticizing the methods of water providers like Northern and Denver Water is akin to marching into a restaurant and telling a chef he is doing everything wrong. I take no pleasure in making life difficult for Brian. I appreciate his love of water history and the knowledge he brings to his job. I am grateful for his devotion to supplying the Front Range with the liquid that allows me to live there. He has helped fill irrigation ditches that have fed me and the people I love. He has brought water to a business in Boulder where a scientist I recently met began working after helping discover the Higgs boson particle—he could bring his brainpower to Boulder because the city has an ample water supply. That water did not magically appear.

It required everything from laborers blasting through mountains with dynamite to people at Northern blasting through environmental blockades to send liquid relief to thirsty farms and cities.

If not for water diversions from the West Slope, my life, and the lives of so many other people on the Front Range, would be impoverished in countless ways. To not acknowledge this, to just open a tap and expect water to flow while criticizing the people who made it so, is disrespectful, if not flat-out crazy.

Greg Silkensen, assistant communications manager for Northern Water, mentioned to me when I was touring Northern's headquarters that he thought a lot of people in the organization were worn out from the constant battles. I told Greg I didn't blame them. Reading about the endless fights over water has pushed me to the edge of exhaustion. The stamina of everyone involved in the conflicts over the upper Colorado River, whether fighting to keep water flowing through its natural channels or to divert it across the Great Divide, is remarkable. I can't imagine going to work every day braced for a new battlefront to open in the unending war to supply the Front Range with water.

Historian Patty Limerick points out in *A Ditch in Time* a common misconception among the public: Water providers have the power to determine population growth. They don't. That power is not in their charters. They are tasked with providing water for the population in their service area. It is up to politicians—and to the public to whom politicians are beholden—to determine a region's carrying capacity, and to then craft policies that tailor population growth to that capacity through land-use planning and zoning laws that encourage density and infill rather than sprawl. If we want to avoid sucking West Slope rivers dry, we should create cities that grow upward rather than outward, making the most efficient use of water for the multitudes moving to Colorado. Water should be priced to reflect both its scarcity and the tremendous costs to communities and to the environment of diverting it from living rivers and drawing it from limited aquifers. And agriculture, which uses the lion's share of the state's water, should continue to increase its irrigation efficiency. To secure the future of the Colorado River headwaters, flooded alfalfa fields and lush green lawns should become vestiges of a vanished past. All of us must become water savvy and make changes to how we use our most valuable resource. The project of saving the Colorado goes far beyond criticizing Northern and Denver Water for their past missteps.

If the men and women of Northern and Denver Water wear black hats, it is because we handed them those hats. We tasked them with providing us water. Then we condemned them for draining rivers to give us the substance we need to stay alive and for our lawns to thrive.

Brian Werner told me a great joke. Like all great jokes, it is sharply truthful. "What's the difference between a developer and an environmentalist in Colorado? The developer is building his house this year. The environmentalist built his house last year."

I have heard Eric Wilkinson, general manager of Northern, referred to by people at odds with Northern's policies on the upper Colorado as "ethical," "principled," and "fair." His job is to deliver water to the Front Range, and he does it well. As does Denver Water's CEO and manager, Jim Lochhead. With a bachelor's degree in environmental biology, he is not a typical water buffalo; he articulates the importance of smart growth in cities to make the most efficient use of water. Ken Neubecker, one of Denver Water's most persistent critics over the past few decades, told me, "I think Jim Lochhead is changing the culture of Denver Water for the better."

The men and women of Northern and Denver Water are driven by a sense of mission to supply water to the drought-plagued plains so farms, homes, businesses, and schools can flourish. Eric Wilkinson and Brian Werner and Jim Lochhead did not set out to damage the ecology of the upper Colorado watershed. And neither did we. But together—Front Range water providers and Front Range water users—we have done precisely that. Even West Slope residents have benefited from draining the Colorado, for the diversions to the Front Range have boosted the state's financial well-being. From providing markets for agricultural products to creating gateways for tourists heading toward the mountains, a thriving Front Range economy is essential to the West Slope's prosperity.

Regardless of the moral ambiguities of providing water and preserving rivers, and regardless of our complicity in the transmountain diversions that have drained the upper Colorado, a bright line must be drawn to safeguard the river's health. We can't throw our hands in the air and say all this is too complicated legally and too complex ethically. We can't give up and accept the status quo while the river dies. Someone who isn't a politician or a bureaucrat, who isn't on anyone's payroll and speaks solely for the Colorado, must say, "This river will not die on my watch."

The murky debate over who is to blame for the river's decline doesn't obscure the clarity of this truth: We are stewards of the river. We have been called upon to deliver it to the next generation better than we found it. Had we started with this principle instead of arriving at it so late, after so many battles, when all of us are so tired, we could have done better by the river. We could have passed it on to the next generation not in a degraded condition but in a state of health. And we could have saved ourselves so much strife, so much fatigue.

In this war of attrition, Bud Isaacs marches on. He will persist until he can see the rocks and the stoneflies in the reach of river for which he feels responsible. He will not stop until the silt has been cleared from the channels that connect him to his grandchildren.

BEFORE BUD BECAME AN oilman he was a ski bum. On weekends in college he headed to Arapahoe Basin to sleep on the patrol shack's floor. After completing avalanche training, he joined the ski patrol. While descending a steep chute one day, sliding snow tumbled him downhill; his training kicked in and he swam for his life, fighting his way toward the light and the air. He had observed safety precautions but the avalanche happened anyway. He says, "You never make a rule with mother nature because she's a fickle whore."

We can build dam after dam to hold back every flood and store each drop, but water breaks our rules. It pushes against each blockade we place in its path, probing for weakness, pressing forward until it finds a way out. Either by dynamite toppling dams or by the slow rot of time, each structure we raise will fall. And every river will, in time, reconnect itself as it travels toward the sea.

Former Secretary of the Interior Bruce Babbitt points out the United States built some 75,000 dams of significant size. This dam-building binge equates to, on average, raising a dam a day since the Declaration of Independence was signed. The one manmade feature on the planet that can be seen from space with the unaided eye is not the Great Wall of China but the green-brown borders that separate irrigated fields from arid lands. A recent study revealed that water depletion has caused the mountains and plains of the West to rise.[31] And scientists have determined that the weight of water stored in the world's reservoirs has sped up the earth's rotation, shifted its axis, and altered the shape of the planet's gravitational field.[32]

Beginning in the Great Depression of the 1930s, we corralled the West's rivers behind dams. Then, as the national mood shifted, our obsession with taming rivers waned. We accepted the unruliness of streams in wild basins filled with wild fish, and we started to celebrate the fickle forces we had worked so hard to control. A movement was born. Momentum to preserve the last of our untamed rivers spread. Soon the rhetoric of "no more dams" ratcheted up to "tear down the dams." And now, in the first decades of the twenty-

31   See "Ongoing Drought-induced Uplift in the Western United States" by Adrian Antal Borsa, Duncan Carr Agnew, and Daniel R. Cayan in the September 26, 2014 issue of the journal *Science*.

32   See "Dams for Water Supply Are Altering Earth's Orbit, Expert Says" by Malcolm W. Browne, *New York Times,* March 3, 1996.

first century, "damolition" is all the rage as a new breed of river keepers replaces the old water buffalos. This generation is as eager to crack apart the concrete walls that hold back rivers as previous generations were to construct them.

Once in the American West the federal government erected mighty structures to keep water's wildness in check, and these engineering marvels partitioned rivers into metered flows. Now, advocates of reconnecting rivers convince the government to blow apart these monumental works. This new guard stands giddy before exploding dams, cheering the blasts of smoke and chunks of shattered concrete. In awestruck wonder they watch the rivers flow as fish swim back upstream. Kilowatt hours of power and rows of irrigated crops are not as important to them as the thrashing tails and hooked jaws of salmon—muscular fish that sniff their way from the ocean to the freshwater streams of their birth.

Salmon are not native to the Colorado River, but the migratory pikeminnow, the so-called "white salmon" of the Colorado, evolved in the river's lower reaches. Similar to the better-known Pacific Northwest salmon, the Colorado pikeminnow has also been devastated by dams. As with the salmon's demise, the pikeminnow's decline could be reversed by destroying structures that have wrecked the connectivity of rivers.

When barriers are removed, the return of the salmon and the pikeminnow to waters where their ancestors evolved is a sign that what has been broken we have healed. Outdated dams across America are being demolished. Rivers are being freed to once again flow as they have for millennia, tumbling down mountains in explosions of spray and foam.

Not all dams should be destroyed, of course. Many serve vital functions of storing the water and providing the power our civilization needs to survive and prosper in the American West. And sometimes the solution to saving a river isn't as simple and exciting as blowing a blockade to smithereens so salmon can swim back upstream. In the case of the ecological crisis for trout and other creatures on the Colorado

River below Windy Gap Dam, the solution isn't destruction. The solution is construction.

Restoring the connectivity of this natural system requires sophisticated engineering. When a bypass that mimics the critical functions of a natural channel is constructed, the sentinel species of giant stoneflies and mottled sculpin will reappear, and the ecosystem's restored health will spread to include trout and dippers, rippling outward to encompass eagles and people.

To anglers who interact with streams, restoring connectivity has intuitive appeal. Citizen scientists with fly rods in hand see rocks and insects tumble downstream; they watch fish swim upstream. They understand that plugging a river with a dam is like blocking a vital artery in a body. If the river is not allowed a pathway to reconnect itself around the dam, it will die. Bypass the blockage and the river has a chance at life. Before scientists made a case for reconnecting the river around Windy Gap, Bud understood the need for a bypass. To a keen observer of nature who is also an engineer trained to solve problems, creating a free-flowing channel around the stagnant reservoir makes obvious sense.

Dave Nickum, who has devoted two decades of his life to protecting coldwater fisheries through his tireless work with Trout Unlimited, told me this about Bud: "More than anyone else, Bud never lost sight of the vision to reconnect the river. He was there on the river before the declines. And as the river went through declines, he was there year by year. He saw it in a very up-close-and-personal way, and that gave him an understanding of what was going on.

"His passion and dedication to a vision of a better Colorado River—trying to make it more like it once was, trying to reconnect it—was always there. Even as whirling disease diminished in importance, Bud still saw the larger picture of the impacts that were going on. He was willing to step up and provide the resources for the research that needed to be done. Through sheer force of will he was able to keep the conversation focused on the question 'How do we get the Colorado River back?'

"A lot of folks, Trout Unlimited included, were focused on flows, habitats—all these other dimensions. Bud would remind us at every opportunity that we've got to reconnect the river—we've got to get the bypass because Windy Gap is creating a dying zone downstream. Bud reminded us that we need to restore the Colorado River to being a river. More than anybody else, he always kept that vision in his mind. He kept bringing other people back to the bypass being an important part of the solution. And in every way he was right.

"The research that Barry Nehring and his colleagues did backed up the fact that problems like sediment movement and water quality were being caused by Windy Gap Dam. Bud was the one who kept the bypass vision alive. And now it's moving closer and closer to becoming a reality."

Chapter 20

THE DEATH AND LIFE OF A GREAT AMERICAN RIVER

N orthern's Windy Gap Firming Project and Denver Water's Moffat Project will put the Colorado River headwaters on a knife-edge of survival. To keep the river from
tipping toward oblivion, the bypass is essential—a growing
consensus confirms this.

For Grand County to grant the 1041 permit allowing
Northern to proceed with its Windy Gap Firming Project, the
Grand County commissioners had to be satisfied that an effective enhancement and mitigation plan was in place. The task of
determining the plan's scientific validity fell on a technical advisory team of diverse stakeholders, including Colorado Parks
and Wildlife, Grand County, Colorado River District, Trout
Unlimited, and Bud's Upper Colorado River Alliance.

The advisory team chose one of five environmental consulting firms that bid on the project. The firm they selected,
Tetra Tech, went to work studying the health of the river, the
potential impacts of the Windy Gap Firming Project, and the
mitigation and enhancement plans. Tetra Tech's *Final Report,
Windy Gap Reservoir Modification Study,* released in February

2015, confirmed what Bud and so many others had been saying all along:

> Aquatic habitat of the Colorado River in the vicinity of
> Windy Gap is sub-optimal. The Gold Medal trout fishery
> appears to be in decline over recent years and fish passage
> is blocked by the dam and impeded in places by shallow
> riffles. Water temperatures commonly exceed standards
> set by the Colorado Department of Public Health and
> Environment during portions of the summer, and recent
> studies suggest key bio-indicator species are nearly extir-
> pated immediately below Windy Gap.

After Tetra Tech published its report, no longer could the
river's declining health be buried or denied. When researchers
conduct independent studies and reach the same conclusion,
the convergence cannot be ignored.

Science is a process of seeking truth. But science can be expensive—and anything that requires money can be corrupted.
A study required for an environmental impact statement is
paid for not by the government but by an organization proposing an action that requires an EIS. Thus, risk of bias always
exists. Organizations with deep pockets can try to purchase
the results they want, and researchers can be convinced to put
aside their principles. Scientists are human and can succumb
to the frailties that afflict us all—greed, ego, willingness to
please. Science, however, purges human error as multiple in-
vestigators scrutinize a problem to arrive at the truth. And the
truth is a beautiful thing—even when it reveals the ugliness of
what's been done to an abused stretch of river.

Grand County's 1041 process led to a collaborative effort
among stakeholders to hire an organization to study the Colorado, eliminating bias. These stakeholders with competing
interests monitored the process through to the final report.
Though Northern footed the bill for the research, many sets of
eyes watched the study unfold.

When the final report was released, it corroborated the
research of Barry Nehring and his colleagues. In layman's

terms, Tetra Tech found that Windy Gap Dam had broken the river, and fixing it required building a bypass around the reservoir. Which, of course, is what Bud had been saying all along.

For many years, Bud and other concerned citizens had watched the river change. Observation is one of our most powerful tools. When we apply logic to the information we gather, we generate hypotheses. After the dam was built, the rocks and bugs in the river disappeared, and so did the fish. It seemed reasonable to conclude the dam was the culprit and a channel to bypass the dam could reverse the damage. Studies supported this.

When scientific research confirms a hypothesis, we can be confident we are moving toward truth—the kind of objective scientific truth that prevents polio from disabling children, sends rovers to Mars, and determines the cause of a river's decline.

BARRY WAS ONCE TOLD by one of his colleagues that sound science always prevails in the end. Barry has built his career, and his life, around the hope that this is the case with Colorado's rivers.

Near Barry's home in Montrose, when I say goodbye to him before heading back to the Front Range, he crushes my hand in his iron grip and chokes back tears as he spins one final tale. He tells me of a former Colorado Division of Wildlife director, Russell George, who allowed the scientists in the department to conduct their research without worrying about political implications or bureaucratic involvement. George tasked field biologists like Barry with doing the best science possible to discern the truth. All that mattered was the evidence the researchers gathered and the logic of their arguments. "That's the way it should be," Barry tells me. "But too many times I've seen people who are worried about politics go out of their way to destroy the credibility of scientists."

Reconnecting the upper Colorado River at Windy Gap will do more than restore populations of insects and fish. It will demonstrate that good science does win in the end.

This victory for the Colorado River will also confirm the value of someone sticking his neck out to protect a resource he cares about. And when that someone is an oilman with the wealth and wherewithal to fight unyieldingly for what he believes is right for the river, the benefits to an embattled natural resource can be profound.

As mentioned, the modern environmental movement in America began in large part with the Sierra Club protesting a proposed dam in Dinosaur National Monument. The Sierra Club rose to prominence in the 1950s by mounting a spirited public information campaign to defend an inviolable principle: no dams in national monuments and national parks. The cancellation of Two Forks Dam in 1990 marked another turning point in water development in the West. Instead of simply saying no to the project, savvy anti-dam crusaders offered viable alternatives to increasing Denver's water supply—and the EPA had little choice but to stop the project, forcing the city and its suburbs to satisfy their water needs by less environmentally destructive means.

When Bud Isaacs combined his passion for preserving the river with his engineer's mindset to find practical solutions, the story of water in the American West took another significant turn. For Bud, a petroleum engineer, and Tony Kay, a software engineer, correcting a problem is more important than pointing out a problem. Bud has always maintained that Windy Gap is a "solvable problem."

For Tony and Bud, the goal has always been clear: Windy Gap Dam must go. They understood that the most important party other than themselves was Northern. Tony explains that Northern's goal has been to get their water. Bud and Tony knew they had to support a solution that made sense for Northern, as well as for the river. The alternative that works for everyone is reconnecting the Colorado with a rebuilt channel. Northern will still be able to make use of its water rights and pump water to Lake Granby for delivery to the Front Range. In the end, everyone wins. Northern's leadership can claim victory in protecting its ability to divert water to Front Range cities and farms. And environmental groups can celebrate the upper Colorado River having a chance to heal after decades of decline.

Tony points out that in this scenario Colorado's Department of Natural Resources also wins. He says, "They can come out of this looking like heroes."

The bypass will create a new stretch of stream open to the public—a revived river channel that will widen children's eyes as they watch trout circle through pools. Small citizens of a state that safeguards a living river, Colorado's children will reach into chilly currents to turn over stones. And they will smell wet riverbanks as they watch dippers dive into the water, these birds entering the teeming kingdom beneath the surface like travelers in a storybook that pass between two worlds.

Tony smiles and says he harbors no hard feelings toward Mike King or anyone else involved in the long-running battle to save the upper Colorado. He is just happy to have a solution that works for everyone and will reverse the damage done to the river. Tony and Bud are looking forward to the future, to leaving the river better than they found it.

The Bureau of Reclamation's Record of Decision on the Windy Gap Firming Project, issued in December 2014, included a bypass—bringing Bud's battle one step closer to completion. Preliminary engineering and design work is underway. Of the approximately $9.6 million needed to construct the bypass, the Colorado Water Conservation Board[33] agreed to pay $2 million, along with the $2 million that Northern's Municipal Subdistrict committed. Bud and Tony now have to find outside contributors to help fund this river restoration

---

33   According to the Colorado Water Conservation Board (CWCB) website, this division of the Colorado Department of Natural Resources "represents each major water basin, Denver and other state agencies in our joint effort to use water wisely and protect our water for future generations."

project. They are securing private money, approaching foundations, and talking with representatives of the federal government.

The U.S. government financed Denver Water's Moffat Tunnel and Northern's Colorado-Big Thompson Project; landowners along the reach of the Colorado harmed by these projects are requesting the federal government help finance the bypass to heal the river. Tony points out that for what amounts to pocket change in the Congressional budget, politicians on both sides of the aisle could score points by restoring an iconic stretch of river and reversing environmental damage caused by federally funded water projects in the Colorado River Basin. "It's an easy bipartisan issue," says Tony.

AS WITH ALL OUR manipulations of nature, management of the Colorado River is an ongoing experiment. Because the effort to repair destruction caused by Windy Gap Dam is based on the best science and engineering, it continues the tradition of the Two Forks Dam victory. And because the fifteen-year struggle has been led by an oil developer, it challenges us enviros to deconstruct the box we've built for ourselves.

The world will never again be a pristine wilderness; it will forevermore be a managed garden. Managing it correctly will require sophisticated planning and engineering. Developers who are demonized by environmentalists often have the financial means and the problem-solving skills to restore natural systems to some semblance of health. All of us use the resources of this planet. And many of us, whether we work for the Sierra Club or extract oil from the earth for a living, want to leave things better than we found them. We want our grandchildren to be startled by a stonefly that seems the size of a sparrow as its silhouette glides across the sky, to hear the small splash of a dipper as it dives into the river, to see the flash of a trout as sunlight glances from its sides.

An enviro's priorities will never align precisely with those of an oil developer. Bud and I butt heads over the scientific consensus that humans are the main driver of climate change, but we agree the river must be saved. As the clamor to preserve the planet grows louder, battles to protect local resources can go unheard. If, in coming decades, humanity averts the searing heat of climate catastrophe but the Colorado is drained to a sterile ditch without sculpin and stoneflies, without dippers and trout, the environmental movement will have partly failed. The book that launched the movement, *Silent Spring,* warned of chemical threats to humanity, but it resonated with readers largely because it told of robins dying in people's backyards. Climate change is a popular meme on social media, but we must also look away from the virtual world on our screens into the real waters of the Colorado. Saving the planet is not enough. We must also care for the rivers that run through our lives.

Our nation once ushered in an age when water projects tamed natural landscapes. Now we can lead the world in stewarding the last scraps of wild nature. What happens at the headwaters of the Colorado River doesn't just affect agriculture, industry, cities, states, and nations. It forms a pivot upon which our species' relationship with the natural world will turn. One of the great environmental battles taking place on the planet is not in some mist-shrouded forest of the Amazon but in our own backyards.

Bud, along with Barry and Tony and so many others, has blown the whistle on an abused stretch of the storied Colorado. These keepers of the river have been criticized and derided, railroaded and ignored. But their observations and studies have not been debunked. The science supports what they've been saying all along.

The river is dying.

Now it's time to restore it to life.

ACKNOWLEDGMENTS

Many people with far more experience and technical expertise than I possess generously contributed their time and talent to make this book possible. I appreciate the following people who made room in their demanding schedules to patiently answer my questions: Brian Werner, Greg Silkensen, Dave Little, Travis Thompson, Rebecca Mitchell, Mike King, Lurline Curran, Ken Kehmeier, Ken Neubecker, Alison Holloran, Gene Reetz, Dan Luecke, Alexandra Davis, Andrew Todd, Patricia Rettig, Jim Lamont, Jim Yust, Chris Garre, Joni Teter, Geoff Elliott, Gary Wockner, Nathan Fey, Nicole Seltzer, Amy Beatie, Greg Hobbs, Jorge Figueroa, Drew Beckwith, Jason Hanson, Woody Beardsley, Tim Nicklas, Ida Sheriff, John Sheriff, Mely Whiting, Kirk Klancke, David Nickum, Steve Bushong, Barry Nehring, Tony Kay, and Bud Isaacs.

I am grateful to Kirk Klancke, Bob Fanch, Fraser Leversedge, and David Nickum for reading manuscript chapters and providing helpful comments. I am especially grateful to Barry Nehring, Tony Kay, Fritz Holleman, Kaye Isaacs, and

Bud Isaacs for scrutinizing the manuscript in its entirety and sharing their feedback.

Several journalists have painstakingly chronicled key developments in the ongoing story of the upper Colorado River. I am particularly indebted to the following: Scott Willoughby, Bruce Finley, Charlie Meyers, Mark Jaffe, Bob Berwyn, and Allen Best.

Any errors, misinterpretations, or oversights are, of course, mine alone.

Thanks to Jason Hanson at Center of the American West for passing this project along to me. Thanks to Courtney Oppel for her diligent proofreading. Many thanks to Allen Jones for thoughtfully shepherding this book through to completion.

Most of all, I am thankful to Bud Isaacs for providing the inspiration and the financial resources to bring this book into the world.

SOURCES

Prologue

American Rivers. "America's Most Endangered Rivers for 2014: Upper Colorado River." www.americanrivers.org/endangered-rivers/2014-report/upper-colorado.

———. *Upper Colorado River Among America's Most Endangered Rivers.* Press release. June 2, 2010. www.americanrivers.org/newsroom/press-releases/upper-colorado-river-among-americas-most-endangered-rivers/#sthash.Dx20XoPB.dpuf.

"California Regulators Adopt Unprecedented Water Restrictions." *National Public Radio,* May 6, 2015. www.npr.org/sections/thetwo-way/2015/05/06/404630607/california-regulators-adopt-unprecedented-water-restrictions.

Clark Howard, Brian. "Historic 'Pulse Flow' Brings Water to Parched Colorado River Delta." *National Geographic,* March 24, 2014.

Cole, Steve, Alan Buis, and Janet Wilson. "Satellite Study Reveals Parched U.S. West Using Up Underground Water." *NASA,* July 24, 2014. www.nasa.gov/press/2014/july/satellite-study-reveals-parched-us-west-using-up-underground-water/#.Vb97KvlViko.

Colorado Foundation for Water Education. "Alarms Sound on the Shrinking Colorado River." *Your Water Colorado Blog,* August 20, 2013. http://blog.yourwatercolorado.org/2013/08/20/alarms-sound-on-the-shrinking-colorado-river.

Deneen, Sally. "Feds Slash Colorado River Release to Historic Lows." *National Geographic,* August 16, 2013. http://news.nationalgeographic.com/news/2013/08/130816-colorado-river-drought-lake-powell-mead-water-scarcity.

"The drying of the West; Drought is forcing westerners to consider wasting less water." *The Economist,* February 22, 2014.

Feeney, Nolan. "Lake Mead Reservoir Hits Record Low." *Time,* June 24, 2015.

Festa, David, and John Entsminger. "A Historic Course Change on the Colorado River." *Las Vegas Review-Journal,* May 29, 2014.

Friends of the Yampa. http://friendsoftheyampa.com.

Jacobsen, Rowan. "The Day We Set the Colorado River Free." *Outside Online,* June 10, 2014. www.outsideonline.com/1928261/day-we-set-colorado-river-free.

Jaffe, Mark. "Current Affairs on State Water; Utilities Cross the Divide to Start Negotiating Water-moving Plans." *The Denver Post,* October 5, 2008.

Kuckro, Rod. "Electricity: Receding Lake Mead Poses Challenges to Hoover Dam's Power Output." *E&E Publishing,* June 30, 2014. www.eenews.net/stories/1060002129.

Leopold, Aldo. "The Green Lagoons" in *A Sand County Almanac: With Essays on Conservation from Round River.* New York: Ballantine Books, 1970.

Miller, Elizabeth. "A River Running; Pulse Flow Feeds More than the Dry Colorado River Delta." *Boulder Weekly,* September 4, 2014.

National Park Service. Grand Canyon. "Geologic Formations." www.nps.gov/grca/learn/nature/geologicformations.htm.

NatureServe Explorer. "Ptycholeilus lucius – Colorado Pikeminnow." http://explorer.natureserve.org/servletNatureServe?search Name=Ptychocheilus+Lucius.

Owen, David. "Where the River Runs Dry." *The New Yorker,* May 25, 2015.

Porter, Eduardo. "The Risks of Cheap Water." *New York Times,* October 14, 2014.

Postel, Sandra. "A Sacred Reunion: The Colorado River Returns to the Sea." Water Currents, *National Geographic,* May 19, 2014. http://voices.nationalgeographic.com/2014/05/19/a-sacred-reunion-the-colorado-river-returns-to-the-sea/.

Powell, James Lawrence. *Grand Canyon: Solving Earth's Grandest Puzzle.* New York: Pi Press, 2005.

Replogle, Jill. "With help from a flood, scientists and activists nurse a bit of the Colorado River Delta back to life." *Public Radio International,* June 23, 2014. www.pri.org/stories/2014-06-23/help-flood-scientists-activists-nurse-bit-colorado-river-delta-back-life.

Rojas-Bracho, Lorenzo, Randall R. Reeves, and Armando Jaramillo-Legorreta. "Conservation of the vaquita Phocoena sinus." *Mammal Review* 36, no. 3 (2006): 179–216.

Sanchez, Ray. "Low California Snowpack Ushers Mandatory Water Restrictions." *CNN,* April 2, 2015. www.cnn.com/2015/04/01/us/california-water-restrictions-drought.

State of California. "California Drought." http://ca.gov/drought.

U.S. Department of the Interior, Bureau of Reclamation, Lower Colorado Region. *Colorado River Basin Water Supply and Demand*

Study. Updated June 2015. www.usbr.gov/lc/region/programs/crbstudy/finalreport.

U.S. Department of the Interior, Bureau of Reclamation, Upper Colorado Region. "Bureau of Reclamation Forecasts Lower Water Release from Lake Powell to Lake Mead for 2014." August 16, 2013. www.usbr.gov/newsroom/newsrelease/detail.cfm?RecordID=44245.

Waterman, Jonathan. "Where the Colorado Runs Dry." *New York Times,* February 14, 2012.

Willoughby, Scott. "Colorado River at Gore Canyon is the stuff dreams are made of." *The Denver Post,* August 31, 2014.

Wines, Michael. "Colorado River Drought Forces a Painful Reckoning for States." *New York Times,* January 5, 2014.

Zielinski, Sarah. "The Colorado River Delta Turned Green after a Historic Water Pulse." *Smithsonian,* December 18, 2014. www.smithsonianmag.com/science-nature/colorado-river-delta-turned-green-after-historic-water-pulse-180953670.

——. "The Colorado River Runs Dry." *Smithsonian,* October 2010. www.smithsonianmag.com/science-nature/the-colorado-river-runs-dry-61427169/.

Chapter 1

Abbey, Edward. *Desert Solitaire: A Season in the Wilderness.* New York: Ballantine Books, 1971.

——. *The Monkey Wrench Gang.* New York: HarperCollins, 2006.

Ault, Toby R., et al. "Assessing the Risk of Persistent Drought Using Climate Model Simulations and Paleoclimate Data." *Journal of Climate* 27, no. 20 (October 2014): 7529–7549.

Bass, Rick. *Oil Notes.* Boston: Houghton Mifflin, 1989.

Beard, Daniel P. *Deadbeat Dams: Why We Should Abolish the U.S. Bureau of Reclamation and Tear Down Glen Canyon Dam.* Boulder, CO: Johnson Books, 2015.

Bower, Tom. *Oil: Money, Politics, and Power in the 21st Century.* New York: Grand Central Publishing, 2009.

Bryce, Robert. *Gusher of Lies: The Dangerous Delusions of Energy Independence.* New York: PublicAffairs, 2009.

——. *Power Hungry: The Myths of "Green" Energy and the Real Fuels of the Future.* New York: PublicAffairs, 2010.

Chouinard, Yvon. "Op-Ed; Tear Down 'Deadbeat' Dams." *New York Times,* May 7, 2014.

Colorado Climate Center. http://climate.colostate.edu.

Cook, Benjamin I., Toby R. Ault, Jason E. Smerdon. "Unprecedented 21st Century Drought Risk in the American Southwest and Central Plains." *Science Advances* 1, no. 1 (February 12, 2015).

Curry, Andrew. "Will Newer Wind Turbines Mean Fewer Bird Deaths?" *National Geographic,* April 28, 2014. http://news.nationalgeographic.com/news/energy/2014/04/140427-altamont-pass-will-newer-wind-turbines-mean-fewer-bird-deaths.

DamNation. Directed by Travis Rummel and Ben Knight. Patagonia, 2014.

Dobb, Edwin. "Alaska's Choice: Salmon or Gold." *National Geographic,* December 2010.

Downey, Morgan. *Oil 101.* New York: Wooden Table Press, 2009.

Fridley, David. "Nine Challenges of Alternative Energy." Excerpted chapter from *The Post Carbon Reader: Managing the 21st Century's Sustainability Crises,* Richard Heinberg and Daniel Lerch, eds. Healdsburg, CA: Watershed Media, 2010.

Friedman, Thomas. *Hot, Flat, and Crowded: Why We Need a Green Revolution—and How It Can Renew America.* New York: Farrar, Straus and Giroux, 2008.

GasLand. Directed by Josh Fox. WOW Company, 2010.

Goodell, Jeff. *Big Coal: The Dirty Secret Behind America's Energy Future.* Boston: Houghton Mifflin Company, 2006.

Graetz, Michael. *The End of Energy: The Unmaking of America's Environment, Security and Independence.* Cambridge, MA: MIT Press, 2011.

Heinberg, Richard. *Powerdown: Options and Actions for a Post-Carbon World.* Gabriola Island, BC: New Society Publishers, 2004.

———. *The Party's Over: Oil, War and the Fate of Industrial Societies.* Gabriola Island, BC: New Society Publishers, 2005.

Hughes, Trevor. "Officials crack down on wind farms that kill birds, bats." *USA Today,* January 29, 2015.

Jacobson, Louis. "Rep. Dan Boren Says Most Domestic Oil Is Produced by 'Small Independent' Companies." *PolitiFact,* May 3, 2011. www.politifact.com/truth-o-meter/statements/2011/may/03/dan-boren/rep-dan-boren-says-most-domestic-oil-produced-smal.

Kaku, Michio. "Solar is the Energy of Tomorrow. Fusion is the Future." *Big Think.* http://bigthink.com/dr-kakus-universe/welcome-to-the-solar-era-followed-by-the-fusion-era.

Kiesecker, Joseph. "Eight Myths and Challenges of Renewable Energy." *The Nature Conservancy Blog,* March 21, 2011. http://blog.nature.org/conservancy/2011/03/21/eigh-myths-and-challenges-of-renewable-energy.

Kunstler, James Howard. *The Long Emergency: Surviving the End of Oil, Climate Change, and Other Converging Catastrophes of the Twenty-First Century.* New York: Grove Press, 2010.

Limerick, Patricia Nelson, with Jason L. Hanson. *A Ditch in Time: The City, the West, and Water.* Golden, CO: Fulcrum, 2012.

Little, Amanda. *Power Trip: From Oil Wells to Solar Cells—Our Ride to the Renewable Future.* New York: HarperCollins, 2009.

Maass, Peter. *Crude World: The Violent Twilight of Oil.* New York: Alfred A. Knopf, 2009.

MacKay, David J. C. *Sustainable Energy—Without the Hot Air.* Cambridge, England: UIT Cambridge Ltd., 2009.

Magill, Bobby. "Colorado's state climatologist says the High Park Fire granted him the permission, courage to talk about climate change." *The Coloradan,* May 23, 2013.

Margonelli, Lisa. *Oil on the Brain: Petroleum's Long Strange Trip to Your Tank.* New York: Broadway Books, 2008.

Maryniak, Gregg. "Storage, Not Generation, is the Challenge to Renewable Energy." *Forbes,* July 20, 2012. www.forbes.com/sites/singularity/2012/07/20/storage-not-generation-is-the-challenge-to-renewable-energy.

McGraw, Seamus. *The End of Country: Dispatches from the Frack Zone.* New York: Random House, 2011.

McPhee, John. *Annals of the Former World.* New York: Farrar, Straus and Giroux, 1998.

National Oceanic and Atmospheric Administration Paleoclimatology Program. "North American Drought: A Paleo Perspective." www.ncdc.noaa.gov/paleo/drought/drght_home.html.

National Renewable Energy Laboratory. www.nrel.gov.

Naugle, David E., ed. *Energy Development and Wildlife Conservation in Western North America.* Washington, DC: Island Press, 2011.

Nuccitelli, Dana. "Fox News found to be a major driving force behind global warming denial." *The Guardian,* August 8, 2013.

Owen, David. *The Conundrum: How Scientific Innovation, Increased Efficiency, and Good Intentions Can Make Our Energy and Climate Problems Worse.* New York: Riverhead Books, 2012.

———. *Green Metropolis: What the City Can Teach the Country about True Sustainability.* New York: Riverhead Books, 2009.

Ozzello, Lori, ed. *Citizen's Guide to Colorado Climate Change.* Denver: Colorado Foundation for Water Education, 2008.

Parry, Simon, and Ed Douglas. "In China, the true cost of Britain's clean, green wind power experiment: Pollution on a disastrous scale." *Daily Mail,* January 26, 2011. www.dailymail.co.uk/home/moslive/article-1350811/In-China-true-cost-Britains-clean-green-wind-power-experiment-Pollution-disastrous-scale.html#.

Pebble Science. http://pebblescience.org.

Pickens, T. Boone. *The First Billion Is the Hardest: Reflections on a Life of Comebacks and America's Energy Future*. New York: Crown Business, 2008.

Powell, James Lawrence. *Dead Pool: Lake Powell, Global Warming, and the Future of Water in the West*. Berkeley: University of California Press, 2008.

Prieto Battery. www.prietobattery.com.

Rocky Mountain Climate Organization. www.rockymountainclimate. org.

Ruppert, Michael. *Confronting Collapse: The Crisis of Energy and Money in a Post Peak Oil World*. White River Junction, VT: Chelsea Green Publishing, 2009.

Seager, Richard. "Persistent Drought in North America: A Climate Modeling and Paleoclimate Perspective." Lamont-Doherty Earth Observatory, The Earth Institute at Columbia University. www. ldeo.columbia.edu/res/div/ocp/drought.

Seinfeld, Jerry. *Comedians in Cars Getting Coffee*. "Howard Stern: The Last Days of Howard Stern." http://comediansincarsgettingcoffee. com/howard-stern-the-last-days-of-howard-stern.

Shepstone, Tom. "There Ought To Be A (Natural Gas) Law—Part I." *Energy In Depth*, April 25, 2012. http://energyindepth.org/ marcellus/there-ought-to-be-a-natural-gas-law-part-i.

Solar Roadways. www.solarroadways.com.

TreeFlow: Streamflow Reconstructions from Tree Rings. www.treeflow. info.

Van Gelder, Lawrence. "Harper Lee Writes Again." *New York Times*, June 28, 2006.

Western Water Assessment. http://wwa.colorado.edu.

Yergin, Daniel. *The Prize: The Epic Quest for Oil, Money and Power*. New York: Free Press, 2009.

———. *The Quest: Energy, Security and the Remaking of the Modern World*. New York, Penguin Books, 2012.

Zielinski, Sarah. "The Western U.S. Could Soon Face the Worst Megadrought in a Millennium." *Smithsonian*, February 12, 2015. www.smithsonianmag.com/science-nature/western-us-could-soon-face-worst-megadrought-millennium-180954238.

Chapter 2

Bass, Rick. *Oil Notes*. Boston: Houghton Mifflin, 1989.

Colorado School of Mines. "Advanced Water Technology Center." http://aqwatec.mines.edu.

Kendall, Trisha Bentz. "First Annual Army ROTC Hall of Fame Ceremony." *Mines: Colorado School of Mines Magazine* 101, no. 1 (Spring 2011).

Marshall, S. L. A. *Battles in the Monsoon: Campaigning in the Central Highlands, Vietnam, Summer 1966.* New York: William Morrow, 1967.

Reisner, Marc. *Cadillac Desert: The American West and Its Disappearing Water.* New York: Penguin, 1993.

Chapter 3

Black, Robert C. *Island in the Rockies: The History of Grand County, Colorado, to 1930.* Boulder: Pruett Publishing Company, 1969.

Bollinger, Edward T. *Rails That Climb: A Narrative History of the Moffat Road.* Golden: Colorado Railroad Museum, 1994.

Central Valley Project. www.usbr.gov/mp/cvp.

Grand County History. "Agriculture of Grand County." http://stories. grandcountyhistory.org/taxonomy/term/38/all?width=80%& height=80%&inline=true.

———. "John & Ida and the Sheriff Ranch." http://stories. grandcountyhistory.org/article/john-ida-and-sheriff-ranch.

Grand Lake Area Historical Society. http://grandlakehistory.org.

Kansas State University. "The ecosystem engineer: Research looks at beavers' role in river restoration." *ScienceDaily,* January 4, 2011. www.sciencedaily.com/releases/2011/01/110103110331.htm.

Mann, Charles C. *1491: New Revelations of the Americas before Columbus.* New York: Knopf, 2005.

Nicklas, Tim. *Winter Park Resort: 75 Years of Imagining More.* Virginia Beach, VA: Donning Company, 2014.

NOAA Habitat Conservation. "Beavers: Mother Nature's First River Restoration Engineers." www.habitat.noaa.gov/highlights/ mothernaturerestorationengineers.html.

Pisani, Donald J. *Water and American Government: The Reclamation Bureau, National Water Policy, and the West, 1902–1935.* Berkeley: University of California Press, 2002.

Pritchett, Lulita Crawford. *Maggie By My Side.* Steamboat Springs, CO: Steamboat Pilot, 1976.

Reisner, Marc. *Cadillac Desert: The American West and Its Disappearing Water.* New York: Penguin, 1993.

Stegner, Wallace. *Beyond the Hundredth Meridian.* New York: Penguin, 1992.

Chapters 5, 6, and 7

Autobee, Robert. *The Colorado-Big Thompson Project*. Denver: Bureau of Reclamation History Program, 1993.

Billington, David P., and Donald C. Jackson. "Big Dams of the New Deal Era: A Confluence of Engineering and Politics." Norman: University of Oklahoma Press, 2006.

Brown, Karla A., ed. *Citizen's Guide to Colorado Water Heritage*. Denver: Colorado Foundation for Water Education, 2004.

———, ed. *Citizen's Guide to Colorado's Environmental Era*. Denver: Colorado Foundation for Water Education, 2005.

Buchholtz, C. W. *Rocky Mountain National Park: A History*. Boulder: Colorado Associated University Press, 1983. www.nps.gov/parkhistory/online_books/romo/buchholtz/chap5.htm.

Carson, Rachel. *Silent Spring*. Boston: Houghton Mifflin, 1987.

Chinatown. Directed by Roman Polanski. Los Angeles: Paramount Pictures, 1974.

Clayton, Jordan A., and Cherie J. Westbrook. "The effect of the Grand Ditch on the abundance of benthic invertebrates in the Colorado River, Rocky Mountain National Park." *River Research and Applications* 24, no. 7 (September 2008): 975–987.

Colorado Water Congress. "Wayne N. Aspinall Award." www.cowatercongress.org/aspinall-award.html.

Colorado Water Conservation Board. "Instream Flow Program." http://cwcb.state.co.us/environment/instream-flow-program/Pages/main.aspx.

———. "Recreational In-Channel Diversions." http://cwcb.state.co.us/environment/recreational-in-channel-diversions/Pages/main.aspx.

Dickman, Pamela. "RMNP to take comment on options for Grand Ditch restoration." *Loveland Reporter-Herald*, March 23, 2012.

Doesken, Nolan, Roger A. Pielke, Sr., and Odilia A. P. Bliss. "Climate of Colorado." Climatography of the United States No. 60, January 2003. http://climate.colostate.edu/climateofcolorado.php.

Edmonds, Carol. *Wayne Aspinall: Mr. Chairman*. Lakewood, CO: Crown Point, 1980.

Erie, Steven P. *Beyond Chinatown: The Metropolitan Water District, Growth, and the Environment in Southern California*. Stanford, CA: Stanford University Press, 2006.

Evans, Jo. *Audubon Rockies Water Handbook: A User-Friendly Guide to Understanding the Basics of Colorado Surface Water Law*. With legal review by Lawrence J. MacDonnell. Fort Collins, CO: Audubon Rockies, 2014.

Getches, David H. *Water Law in a Nutshell*. St. Paul, MN: Thompson West, 2009.

Grace, Stephen. *Dam Nation: How Water Shaped the West and Will Determine Its Future*. Guilford, CT: Globe Pequot Press, 2012.

——. *The Great Divide*. Guilford, CT: TwoDot, 2015.

Grand County History. "Water from the Mountains—The Grand Ditch." http://stories.grandcountyhistory.org/category/water.

Hansen, James E. *The Water Supply & Storage Company: A Century of Colorado Reclamation, 1891–1991*. Fort Collins, CO: The Company, 1991.

Harvey, Mark W. T. *A Symbol of Wilderness: Echo Park and the American Conservation Movement*. Albuquerque: University of New Mexico Press, 1994.

Hiltzik, Michael. *Colossus: Hoover Dam and the Making of the American Century*. New York: Free Press, 2010.

Hinchman, Steve. "EPA to Denver: Wake up and Smell the Coffee!" *High Country News*, April 10, 1989.

Hobbs, Gregory J., Jr. *Citizen's Guide to Colorado Water Law*. 3rd ed. Denver: Colorado Foundation for Water Education, 2009.

——. *The Public's Water Resource: Articles on Water Law, History, and Culture*. Denver: Continuing Legal Education in Colorado, 2007.

Hundley, Norris, Jr. *Water and the West: The Colorado River Compact and the Politics of Water in the American West*. Berkeley: University of California Press, 2009.

Jones, P. Andrew, and Thomas Cech. *Colorado Water Law for Non-Lawyers*. Boulder: University Press of Colorado, 2009.

Kahrl, William. *Water and Power: The Conflict Over Los Angeles' Water Supply in the Owens Valley*. Berkeley: University of California Press, 1982.

Los Angeles Department of Water and Power. "The Story of the Los Angeles Aqueduct." http://wsoweb.ladwp.com/Aqueduct/historyoflaa/index.htm.

MacDonnell, Lawrence J. *From Reclamation to Sustainability: Water, Agriculture, and the Environment in the American West*. Niwot: University Press of Colorado, 1999.

Martin, Russell. *A Story that Stands Like a Dam: Glen Canyon and the Struggle for the Soul of the West*. New York: Henry Holt & Co., 1989.

McPhee, John. *Encounters with the Archdruid*. New York: Farrar, Straus and Giroux, 1971.

Muir, John. "The Hetch Hetchy Valley." *Sierra Club Bulletin* 6, no. 4 (January 1908).

Mulholland, Catherine. *William Mulholland and the Rise of Los Angeles*. Berkeley: University of California Press, 2002.

Northern Water. "Colorado-Big Thompson Project." www.northern water.org/WaterProjects/C-BTProject.aspx.

Obmascik, Mark. "Dam option offered: Kinder, cheaper Animas-La Plata?" *The Denver Post*, October 12, 1995.

——. "EPA aide urges death sentence for Two Forks; Final word on controversial project up to national water." *The Denver Post*, March 27, 1990.

——. "EPA chief adamant on Two Forks; Reilly offers no encouragement to Colo. officials about dam's future." *The Denver Post*, March 17, 1990.

——. "EPA holds final Two Forks forums; Backers, foes mount huge grassroots campaigns to pack public hearings." *The Denver Post*, October 23, 1989.

——. "EPA voices objections to Animas-La Plata." *The Denver Post*, December 17, 1992.

——. "Two Forks lauded and denounced: Noisy crowd gathers at hearing over federal plans to kill dam project." *The Denver Post*, October 24, 1989.

Pisani, Donald J. *Water and American Government: The Reclamation Bureau, National Water Policy, and the West, 1902–1935*. Berkeley: University of California Press, 2002.

Powell, James Lawrence. *Dead Pool: Lake Powell, Global Warming, and the Future of Water in the West*. Berkeley: University of California Press, 2008.

Reisner, Marc. *Cadillac Desert: The American West and Its Disappearing Water*. New York: Penguin, 1993.

Satchell, Michael. "The Last Water Fight: A Dam That Won't Die Shows Power of Pork." *U.S. News & World Report*, October 23, 1995.

Schulte, Steven C. *Wayne Aspinall and the Shaping of the American West*. Boulder: University Press of Colorado, 2002.

Sibley, George. *Water Wranglers: The History of the Colorado River District: A Story about the Embattled Colorado River and the Growth of the West*. Grand Junction: Colorado River District, 2012.

Smythe, William E. *The Conquest of Arid America*. New York: Macmillan, 1905.

Stegner, Wallace. *Beyond the Hundredth Meridian*. New York: Penguin, 1992.

Stenzel, Richard, and Tom Cech. *Water, Colorado's Real Gold: A History of the Development of Colorado's Water, The Prior Appropriation Doctrine and the Colorado Division of Water Resources*. Denver: Richard Stenzel, 2013.

Sturgeon, Stephen C. *The Politics of Western Water: The Congressional Career of Wayne Aspinall*. Tucson: University of Arizona Press, 2002.

Tyler, Daniel. *The Last Water Hole in the West: The Colorado-Big Thompson Project and the Northern Colorado Water Conservancy District.* Niwot: University Press of Colorado, 1992.

———. *Silver Fox of the Rockies: Delphus E. Carpenter and Western Water Compacts.* Norman: University of Oklahoma Press, 2003.

Tyson, Neil deGrasse. *Space Chronicles: Facing the Ultimate Frontier.* New York: W.W. Norton, 2012.

U.S. Congress. Senate. 75th Congress, 1st Session. *Document No. 80. Synopsis of Report on Colorado-Big Thompson Project, Plan of Development and Cost Estimate Prepared by the Bureau of Reclamation, Department of the Interior.* June 15, 1937.

U.S. Department of the Interior, Bureau of Reclamation. "Colorado-Big Thompson Project." www.usbr.gov/projects/Project.jsp?proj_Name=Colorado-Big+Thompson+Project.

Visty, Judy. "Effects of the Grand Ditch." National Park Service, U.S. Department of the Interior, September 2004. www.nps.gov/romo/learn/management/upload/ditch.pdf.

Wilkinson, Charles. *Crossing the Next Meridian: Land, Water, and the Future of the West.* Washington, DC: Island Press, 1992.

Worster, Donald. *Rivers of Empire: Water, Aridity and the Growth of the American West.* New York: Pantheon, 1985.

Chapter 8

Archer, D. L., et al. *A study of the endangered fishes of the upper Colorado River.* Grand Junction: Fish and Wildlife Service, Colorado River Fishery Project, 1985.

Best, Allen. "Collaboration, but not all cards were dealt equally." *Mountain Town News,* April 18, 2011. http://mountaintownnews.net/2011/04/18/collaboration-but-not-all-cards-were-dealt-equally/.

Brown, Karla A., ed. *Citizen's Guide to Colorado's Environmental Era.* Denver: Colorado Foundation for Water Education, 2005.

Colorado River Basin Salinity Control Forum. http://coloradoriversalinity.org.

Environmental Protection Agency. "EPA History: National Environmental Policy Act." www.epa.gov/nepa.

Haldane, J. B. S. *Possible Worlds and Other Essays.* London: Chatto and Windus, 1932.

Marston, Ed. "Water Pressure." *High Country News,* November 20, 2000.

Northern Water. "How Windy Gap Works." www.northernwater.org/WaterProjects/HowWindyGapWorks.aspx.

———. "Windy Gap Project History." www.northernwater.org/About Us/WindyGapHistory.aspx.

Silkensen, Gregory M. *Technical Report No. 61. Windy Gap: Transmountain Water Diversion and the Environmental Movement.* Colorado Water Resources Research Institute, Colorado State University, August 1994.

Tyler, Daniel. *The Last Water Hole in the West: The Colorado-Big Thompson Project and the Northern Colorado Water Conservancy District.* Niwot: University Press of Colorado, 1992.

Chapters 9 and 10

2010–2011 Grand County Resource Guide. Winter Park, CO: Guest Guide Publications.

Abrams, Dan. "The Worth of a Wild Trout." *Wild Trout III, Proceedings of the Symposium*, Yellowstone National Park, September 24–25, 1984.

Bartholomew, Jerri L., and Paul W. Reno. "The history and dissemination of whirling disease." *American Fisheries Society Symposium* 29, Bethesda, MD: 2002.

Behnke, Robert J. *Trout and Salmon of North America.* New York: Free Press, 2002.

Blumhardt, Miles. "Something fishy: Whirling disease may be taking toll on trout fry in rivers." *Fort Collins Coloradoan*, October 23, 1994.

Colorado Parks and Wildlife. "Whirling Disease and Colorado's Trout." http://cpw.state.co.us/learn/Pages/WhirlingDiseaseCOTrout.aspx.

Ewert, John. *Colorado River near Parshall, Fish Survey and Management Information.* Colorado Parks and Wildlife, 2013.

Halverson, Anders. *An Entirely Synthetic Fish: How Rainbow Trout Beguiled America and Overran the World.* New Haven, CT: Yale University Press, 2010.

Hinchman, Steve. "How Dam Opponents Developed and Refined a Strategy." *High Country News*, April 10, 1989.

Hinshaw, Glen. *Crusaders for Wildlife: A History of Wildlife Stewardship in Southwestern Colorado.* Lake City, CO: Western Reflections Publishing Company, 2000.

Kafka, Franz. *The Trial.* New York: Schocken Books, 1998.

Limerick, Patricia Nelson, with Jason L. Hanson. *A Ditch in Time: The City, the West, and Water.* Golden, CO: Fulcrum, 2012.

Meyers, Charlie. "Colorado: A trout river of no return?" *The Denver Post*, July 27, 1997.

———. "Creating a river bypass might be the solution." *The Denver Post,* January 1, 2002.

———. "CWC ruling on fishing policies approaches trout creel limit, whirling disease crux of Thursday's public debate." *The Denver Post,* November 12, 2000.

———. "Diseased Trout Remain on Hook for Late Reprieve." *The Denver Post,* April 30, 1995.

———. "DOW Loses Activist as Kochman Set to Retire." *The Denver Post,* April 7, 2002.

———. "DOW, Trout Unlimited Feuding." *The Denver Post,* October 24, 1999.

———. "Fish Stocking Policy Will Hamper Anglers." *The Denver Post,* November 13, 2001.

———. "Fisherman Evolve as Stocking Shifts; Fingerlings Set Changes." *The Denver Post,* June 1, 2003.

———. "Heralded rainbows battle for survival; whirling disease threatening fish in prime Colorado River waters." *The Denver Post,* August 7, 1994.

———. "Low Fish Stocks, Sales Put DOW in Dilemma." *The Denver Post,* September 16, 2001.

———. "Myopia cropping into disease's debate; Trout strain given undue write-offs." *The Denver Post,* May 26, 2002.

———. "Number of Trout-Loving Anglers Doesn't Add Up." *The Denver Post,* August 14, 2001.

———. "Rainbow trout questions yield no easy answers." *The Denver Post,* October 4, 1994.

———. "Reeling in evidence for an answer; Nehring has line on fatal fish ailment." *The Denver Post,* March 4, 2001.

———. "Spotlight is now on the brown as rainbow numbers keep on dwindling." *The Denver Post,* February 28, 2001.

———. "State's Trout Program Can't Be a Meat Market." *The Denver Post,* March 15, 1998.

———. "Stocking Policy Hurting Spread of 'Clean' Trout." *The Denver Post,* February 17, 2004.

———. "Uncertainties Cloudy Hatchery System Future." *The Denver Post,* March 18, 2001.

———. "Whirling disease is with us." *The Denver Post,* September 9, 1997.

Nehring, R. Barry. "Whirling Disease in Feral Trout Populations in Colorado." E. P. Bergersen and B. A. Knopf, eds. *Whirling Disease Workshop Proceedings: Where do we go from here?* Fort Collins: Colorado Cooperative Fish and Wildlife Research Unit, 1996: 126–144.

——. *Biological, environmental, and epidemiological evidence implicating Myxobolus cerebralis (the protozoan parasite causing whirling disease) in the dramatic decline of the wild rainbow trout population in the Colorado River in Middle Park, Colorado.* Colorado Division of Wildlife, 1993.

——, et al. "Impacts of whirling disease on wild rainbow trout in the South Platte River." *Whirling Disease Symposium: Expanding the Database.* Whirling Disease Foundation, 1997: 1–9.

——, et al. "Using Sediment Core Samples to Examine the Spatial Distribution of Myxobolus cerebralis Triactinomyxons in Windy Gap Reservoir, Colorado." *North American Journal of Fisheries Management* 23 (2003): 376–384.

——, et al. "Laboratory Studies Indicating that Living Brown Trout *Salmo trutta* Expel Viable *Myxobolus cerebralis* Myxospores." *American Fisheries Society Symposium* 29, Bethesda, MD: 2002.

——, and Kevin G. Thompson. *Evaluating the Relative Abundance of Strains of* Tubifex tubifex *With Varying Vulnerability to* Myxobolus cerebralis *in Windy Gap Reservoir, Colorado Before and After a Dramatic Decline in Actinospore (TAM) Production.* Colorado Division of Wildlife, 2005.

——, and Kevin G. Thompson. *Special Report Number 76. Impact Assessment of Some Physical and Biological Factors in the Whirling Disease Epizootic among Wild Trout in Colorado.* Colorado Division of Aquatic Research, March 2001.

——, Kevin G. Thompson, and Michael Catanese. *Decline in TAM Production in Windy Gap Reservoir: Is It Linked To a Major Shift in Relative Abundance of Tubificid Lineages?* Colorado Division of Wildlife, 2005.

——, and Peter G. Walker. "Whirling disease in the wild: The new reality in the Intermountain West. *Fisheries* 21, no. 6 (1996): 28–30.

Nickum, David. *Whirling Disease in the United States: Overview and Guidance for Research and Management.* Trout Unlimited, March 1996. http://whirlingdisease.montana.edu/pdfs/TU_Summary_1996.pdf.

——. *Whirling Disease in the United States: A Summary of Progress in Research and Management.* Trout Unlimited, January 1999. http://whirlingdisease.montana.edu/pdfs/TU_Report_99.pdf.

Obmascik, Mark. "Dam within a dam proposed; Second at reservoir hoped to halt spread of whirling disease." *The Denver Post,* March 15, 1999.

——. "EPA holds final Two Forks forums; Backers, foes mount huge grassroots campaigns to pack public hearings." *The Denver Post,* October 23, 1989.

———. "Spawning Disaster; Whirling disease kills trout, has no cure."
The Denver Post, April 16, 1995.

———. "Tide turns on whirling disease; Spring water helping state make
progress against trout scourge." *The Denver Post,* May 29, 1999.

———. "Trout-disease toll rising in state rivers." *The Denver Post,*
December 28, 1998.

———. "Trout quarantine pursued in legislature." *The Denver Post,*
April 20, 1995.

———. "Two Forks lauded and denounced: Noisy crowd gathers at
hearing over federal plans to kill dam project." *The Denver Post,*
October 24, 1989.

Preston, Richard. *The Hot Zone: The Terrifying True Story of the Origins
of the Ebola Virus.* New York: Anchor Books, 1995.

Saile, Bob. "Documented studies show WD is taking toll on trout." *The
Denver Post,* July 25, 1995.

———. "DOW calls moratorium on stream-stocking plans." *The Denver
Post,* April 16, 1995.

———. "Nehring, Walker honored for whirling-disease work." *The
Denver Post,* April 23, 1996.

———. "Salmon, trout at risk of disease." *The Denver Post,* April 8, 1994.

———. "Stocking policy is up for review." *The Denver Post,* March 7,
1995.

———. "Tainted trout feared at U.S. hatchery: Whirling disease
apparently spreads from reservoir near Leadville." *The Denver Post,*
April 1, 1995.

———. "What's at end of rainbow? Whirling disease requires serious
studies." *The Denver Post,* May 1, 1994.

———. "Whirling disease affecting DOW." *The Denver Post,* April 17,
1994.

———. "Whirling disease fears weren't so reactionary." *The Denver Post,*
December 11, 1994.

———. "Whirling disease in states of denial." *The Denver Post,* February
19, 1995.

———. "Whirling disease on the rise; Fish parasite takes ride on wild
side." *The Denver Post,* January 8, 1995.

———. "Whirling disease takes toll; Wild rainbow trout dying by
thousands." *The Denver Post,* December 8, 1994.

Soulé, Michael E., and Bruce A. Wilcox, eds. *Conservation Biology:
An Evolutionary-Ecological Perspective.* Sunderland, MA: Sinauer
Associates, 1980.

Walker, Peter G., and R. Barry Nehring. *An investigation to determine
the cause(s) of the disappearance of young wild rainbow trout in the
upper Colorado River, in Middle Park, Colorado.* Colorado Division
of Wildlife, 1995.

Weller, Robert. "Disease may have spread to hardier brown trout." *The Associated Press,* June 16, 1996.

Whirling Disease Initiative. http://whirlingdisease.montana.edu/about/faq.htm.

Wood, Chris. "Don't Blame Anglers." *Trout Unlimited Blog,* April 17, 2015. hwww.tu.org/blog-posts/dont-blame-anglers.

Chapters 11 and 12

2010–2011 Grand County Resource Guide. Winter Park, CO: Guest Guide Publications.

Adams, Susan B., and David A. Schmetterling. "Freshwater sculpins: phylogenics to ecology." *Transactions of the American Fisheries Society* 136, no 6. (2007): 1736–1741.

Articles of Incorporation of the Upper Colorado River Alliance. May 2011.

Carson, Rachel. *The Sense of Wonder.* New York: Harper & Row, 1965.

Crowl, Douglas. "Big Rewards or Big Impact? More Time Given to Chimney Hollow Proposal." *Loveland Connection,* October 16, 2008.

Culver, Virginia. "Colorado developer George Beardsley dies at 74; had environmental bent." *The Denver Post,* August 16, 2011.

Discover John Muir. http://discoverjohnmuir.com.

Erickson, Robert C. *Benthic Field Studies for the Windy Gap Study Reach, Colorado River, Colorado, Fall 1980 to Fall 1981.* Report prepared for Northern Colorado Water Conservancy District, Municipal Sub-District, 1983.

Finley, Bruce. "Greenback cutthroat trout recovery gets stronger with Colorado project." *The Denver Post,* August 28, 2014.

Leopold, Aldo. *Round River: From the Journals of Aldo Leopold.* New York: Oxford University Press, 1953.

Nature's Notebook. "Pteronarcys californica." www.usanpn.org/nn/Pteronarcys_californica.

Nehring, Barry R. *Colorado River Aquatic Resource Investigations, Federal Aid Project F-273-R19.* Colorado Parks and Wildlife, August 2012.

——, Brian Heinold, and Justin Pomeranz. *Job 1: Colorado River Aquatic Invertebrate Investigations. Colorado River Aquatic Resources Investigations, Federal Aid Project F-237R-18.* Colorado Division of Wildlife, June 2011.

——, Brian Heinold, and Justin Pomeranz. *Job 2: Colorado River Mottled Sculpin Population Studies. Colorado River Aquatic Resources Investigations, Federal Aid Project F-237R-18.* Colorado Division of Wildlife, June 2011.

———, Jon Ewert, and Sherman Hebein. *A Review of Aquatic Invertebrate Studies and Fish Population Survey Data for the Colorado River in Middle Park, Colorado from 1980 through 2009: What Does It Tell Us?* Colorado Division of Wildlife, January 2010.

Northern Arizona University. Land Use History of the Colorado Plateau. "Biotic Communities of the Colorado Plateau, Riparian Areas." http://cpluhna.nau.edu/Biota/riparian_communities.htm.

Northern Water. "Windy Gap Firming Project." www.northernwater. org/WaterProjects/WGFProjectOverview.aspx.

Resource Engineering, Inc. Denver Moffat Project. Presentation to the Wildlife Commission on Behalf of the Upper Colorado River Alliance. Glenwood Springs, Colorado, May 6, 2011.

Sahagun, Louis. "In Owens Valley, Water Again Flows." *Los Angeles Times,* December 7, 2006.

———. "River Is Resurrected." *Los Angeles Times,* July 8, 2007.

Trout Unlimited. "Windy Gap Firming Project." http://coloradotu.org/ windy-gap.

U.S. Department of the Interior, Bureau of Reclamation. *Principles to Govern the Release of Water at Granby Dam to Provide Fishery Flows Immediately Downstream in the Colorado River.* 1961.

———. *Windy Gap Firming Project Final Environmental Impact Statement.* November 2011. www.usbr.gov/gp/ecao/wgfp_feis.

Chapter 13

Beckwith, Drew, et al. *Filling the Gap: Commonsense Solutions for Meeting Front Range Water Needs.* Boulder: Western Resource Advocates, Trout Unlimited, and the Colorado Environmental Coalition, 2011. www.westernresourceadvocates.org/publications/ filling-the-gap-front-range/.

Buchanan, Dave. "Wildlife Commission approves mitigation plans for diversions on upper Colorado River." *The Daily Sentinel,* June 11, 2011.

Bushong, Steve. "Windy Gap and Moffat Tunnel Firming Projects." Letter to Christine Scanlon, Senior Staff/Director of Legislative Affairs; Doug Young, Senior Policy Director; Governor John W. Hickenlooper. May 27, 2011.

Colorado Water Conservation Board. "37-60-122.2. Fish and wildlife resources - legislative declaration - fish and wildlife resources fund - authorization." http://cwcb.state.co.us/legal/Documents/ Statutes/37-60-122_2.pdf.

Colorado Wildlife Federation. "Windy Gap Firming Project Update." February 22, 2012. http://coloradowildlife.org/OUR-STAND/ WINDY-GAP-FIRMING-PROJECT-UPDATE.HTML.

Denver Water. *Moffat Collection System Project Fish and Wildlife Mitigation Plan.* June 9, 2011.

———, and CDM Smith. *Moffat Collection System Project Request for Clean Water Act Section 401 Water Quality Certification, Final Report.* June 2015. www.colorado.gov/pacific/sites/default/files/DenverWaterMoffat%20401_06-30-2015_Final.pdf.

Environmental Protection Agency. Comments on Windy Gap Firming Project Draft Environmental Impact Statement, December 19, 2008.

———. Comments on Windy Gap Firming Project Final Environmental Impact Statement; CEQ # 20110413, February 6, 2012. www.defendthecolorado.org/sites/default/files/WGFP%20FEIS%20Final%2002-06-2012.pdf.

Finley, Bruce. "EPA Wants Further Review of Water-diversion Project to Protect Colorado River." *The Denver Post,* February 17, 2012.

Holleman, Fritz. "Moffat and Windy Gap Firming Draft Mitigation Plans." Letter to Mike King, Director Colorado Division of Wildlife. February 24, 2011.

Horn, John. "Patagonia's new line of activism is documentary 'DamNation.'" *Los Angeles Times,* May 1, 2014.

Isaacs, Bud. Letter to Tom Remington, Director of Colorado DOW. May 19, 2011.

———. "Saving the Colorado River." Letter to the Editor. *The Denver Post,* March 31, 2012.

Jaffe, Mark. "Surge of water projects show little coordination." *The Denver Post,* April 26, 2009.

Kurbjun, Janice. "Is the Colorado Dying?" *Summit Daily News,* June 12, 2011.

Municipal Subdistrict Northern Colorado Water Conservancy District. *Windy Gap Firming Project Fish and Wildlife Mitigation Plan.* June 9, 2011. www.northernwater.org/docs/windygapfirming/wgfpmitigationplanjune2011.pdf.

Nehring, Barry R., Brian Heinold, and Justin Pomeranz. *Colorado River Aquatic Resources Investigations – Federal Aid Project F-237R-18.* Draft Report. Colorado Division of Wildlife, June 2011.

Peternell, Drew. "Sucking the river dry." *The Denver Post,* April 8, 2012.

U.S. Department of the Interior, Bureau of Reclamation. *Windy Gap Firming Project Final Environmental Impact Statement.* www.usbr.gov/gp/ecao/wgfp_feis.

Whiting, Mely. Counsel, Colorado Water Project. Letter to Colorado Wildlife Commission. Trout Unlimited Comments – Fish and Wildlife Mitigation Recommendations, Windy Gap Firming Project and Moffat Tunnel Collection System Project, May 26, 2011.

Willoughby, Scott. "Concerns remain over approved water projects." *The Denver Post*, June 12, 2011.

———. "Preservation facing pivotal vote." *The Denver Post*, June 8, 2011.

———. "River advocates fear that two proposed water diversions could hurt fishing on the upper Colorado." *The Denver Post*, March 21, 2012.

———. "Water is Rising on Hickenlooper over Diversion Plan." *The Denver Post*, February 22, 2012.

———. "Windy Gap Remains a Controversial Topic." *The Denver Post*, August 4, 2012.

Chapters 14, 15, 16, and 17

American Rivers. "America's Most Endangered Rivers of 2005." www.americanrivers.org/wp-content/uploads/2013/10/mer_2005.pdf?bf5c96.

Beckwith, Drew, and Dan Luecke. *Filling the Gap: Commonsense Solutions for Meeting Front Range Water Needs.* Western Resource Advocates, Trout Unlimited and Colorado Environmental Coalition, February 2011.

Berwyn, Bob. "Colorado: Forest Service comment letter shows breadth and depth of impacts from Denver Water's diversion plan." *Summit County Citizens Voice*, June 23, 2014. http://summitcountyvoice.com/tag/moffat-tunnel-collection-system-expansion.

———. "Colorado: Proposed Water Deal Could End Decades of Fighting." *Summit County Citizens Voice*, April 28, 2011. http://summitcountyvoice.com/2011/04/28/colorado-proposed-water-deal-could-end-decades-of-fighting.

———. "Kumbaya on the Colorado River?" *Summit County Citizens Voice*, April 27, 2011. http://summitcountyvoice.com/2011/04/27/kumbaya-on-the-colorado-river.

———. "Proposed water diversion and Gross Reservoir expansion may trigger all-out water war." *Boulder Weekly*, May 1, 2014.

Best, Allen. "2002 drought found to be worst in 300 years." *Summit Daily*, January 4, 2004.

———. "Collaboration, but not all cards were dealt equally." *Mountain Town News*, April 18, 2011. http://mountaintownnews.net/2011/04/18/collaboration-but-not-all-cards-were-dealt-equally.

———. "Dillion Dam, water rights, and the shouting match of the '50s." *Mountain Town News*, September 8, 2013. http://mountaintownnews.net/2013/09/08/dillon-dam-and-shouts-in-the-past.

Bina, Tonya. "History in the making, water agreement touted as game changer for Colorado Water Management." *Ski-Hi News,* May 15, 2012.

Bledsoe, Brian P., and Johannes Beeby. *Sedimentation processes and effects in the Fraser River and its tributaries.* Report prepared for Trout Unlimited, June 22, 2012.

———, Johannes Beeby, and K.W. Hardie. *Evaluation of Flushing Flows in the Fraser River and Its Tributaries.* Colorado Trout Unlimited, September, 2013.

Bollinger, Edward T. *Rails That Climb: A Narrative History of the Moffat Road.* Golden: Colorado Railroad Museum, 1994.

Bounds, Amy. "Boulder County refuses to sign Gross Reservoir agreement." *Boulder Daily Camera,* January 7, 2013.

Brady, Kristyn. "Heroes of Conservation, 2011 Winner, Kirk Klancke." *Field & Stream.* www.fieldandstream.com/heroes/conservation/finalists/kirk-klancke.

Braet, Megan. *A Dry Legacy: The Challenge for Colorado's Rivers.* Trout Unlimited and Colorado Water Project, January 2002.

Brennan, Charlie. "EPA: Plan Threatens Water Quality." *Boulder Daily Camera,* June 19, 2014.

Bublitz, Bill. "Statue of Dwight Eisenhower on the Fraser River to be unveiled Saturday." *Ski-Hi News,* August 7, 2008.

Casamassa, Glen P. Forest Supervisor, U.S. Department of Agriculture, Forest Service. Letter to Rena Brand, Denver Regulatory Office, U.S. Army Corps of Engineers. June 9, 2014. www.scribd.com/doc/230903178/USFS-Moffat-FEIS-Comment-Letter.

Coley/Forrest, Inc. *Water and Its Relationship to the Economies of the Headwaters Counties.* December 2011.

Colorado Business Hall of Fame. "Cathey M. Finlon." www.colorado businesshalloffame.org/cathey-m-finlon.html.

Colorado Headwaters Chapter of Trout Unlimited. www.coheadwaters.org.

Colorado: The Official State Web Portal, Gov. John Hickenlooper. "Historic Proposed Agreement Heralds a New Path for Colorado Water." www.colorado.gov/cs/Satellite/GovHickenlooper/CBON/1251592084762.

"Denver Goes to West Slope for Additional Water Supply." *Engineering News Record* 115 (July–December 1935).

The Denver Post Editorial Board. "The Future of Gross Reservoir." *The Denver Post,* January 11, 2013.

Denver Water. "A Balanced Approach." Mile-High Water Talk, August 13, 2013. http://denverwaterblog.org/2013/08/13/a-balanced-approach.

———. "Colorado River Cooperative Agreement: Path to a Secure Water Future." www.denverwater.org/SupplyPlanning/Planning/ColoradoRiverCooperativeAgreement.

———. "Conservation." www.denverwater.org/conservation.

———. "History of the Northern Collection System." www.denverwater.org/AboutUs/History/NorthernCollectionSystem.

———. "History of the South Platte Collection System." www.denverwater.org/AboutUs/History/SouthPlatteCollection.

———. "Moffat Collection System Project." www.denverwater.org/supplyplanning/watersupplyprojects/moffat.

———. "Service Area." www.denverwater.org/AboutUs/ServiceArea.

Devil's Thumb Ranch. www.devilsthumbranch.com.

Dye, Homer Jr. "A Great Dream Opens an Empire." *Popular Science Monthly,* June 1928.

Eha, Walter. *Water for Denver: How Water Helped Build a City.* Denver: Denver Water/Denver Public Library, 2006.

Ehrlich, Paul R. *The Population Bomb.* New York: Ballantine Books, 1971.

The Environmental Group. "Top Ten fatal flaws with Moffat/Gross." http://tegcolorado.org/top-ten-fatal-flaws-with-moffat-gross.html.

Ewert, John. "Fraser River: Safeway." *Colorado Headwaters Fisheries Management,* March 17, 2013. http://coloradoheadwatersfisheries.blogspot.com.

Finley, Bruce. "Agencies agree to tackle problem of traction-sand deposits in Fraser River." *The Denver Post,* November 15, 2010.

Forrest, Kenton, and Charles Albi. *The Moffat Tunnel: A Brief History.* Golden: Colorado Railroad Museum, 1985.

Front Range Pika Project. www.pikapartners.org/cwis438/websites/FRPP/Home.php?WebSiteID=18.

Gardner-Smith, Brent. "Colorado's instream flow program is lauded, challenged." *Aspen Daily News,* January 21, 2014.

———. "Water Group: Look Elsewhere for Water." *Aspen Times,* December 2, 2013.

Gilboy, Cecelia. "Obituary for a River." *Boulder Weekly,* June 26, 2014.

Grand County. "Colorado River Cooperative Agreement." www.co.grand.co.us/416/Colorado-River-Cooperative-Agreement-CRC.

———. "Fraser River Sediment Project." www.co.grand.co.us/411/Fraser-River-Sediment-Project.

———. "Grand County Stream Management Plan, Phase 3." www.co.grand.co.us/412/2462/Stream-Management-Plan-Phase-3.

———. Summit County, Northwest Colorado Council of Governments, Middle Park Water Conservancy District, Trout Unlimited, Colorado River Water Conservation District, Western Resource

Advocates. *Moffat Collection System Project Draft Environmental Impact Statement Joint Rebuttal Report.* March 17, 2010.

Hestmark, Martin. Assistant Regional Administrator, U.S. Environmental Protection Agency, Region 8. "Moffat Collection System Project Final Environmental Impact Statement, CEQ #20140129." Letter to Rena Brand, Denver Regulatory Office, U.S. Army Corps of Engineers. June 9, 2014. www.scribd.com/doc/230416366/EPA-Moffat-Comments.

Hinchman, Steve. "EPA to Denver: Wake Up and Smell the Coffee." *High Country News,* April 10, 1989.

Leonard, Stephen, and Thomas Noel. *Denver: Mining Camp to Metropolis.* Niwot, CO: 1990.

Limerick, Patricia Nelson, with Jason L. Hanson. *A Ditch in Time: The City, the West, and Water.* Golden, CO: Fulcrum, 2012.

Magill, Bobby. "Colorado's state climatologist says the High Park Fire granted him the permission, courage to talk about climate change." *The Coloradan,* May 23, 2013.

McCarthy, Cormac. *The Stonemason: A Play in Five Acts.* Hopewell, NJ: Ecco Press, 1994.

Meyer, Jeremy P. "Hayman fire still mucking up water." *The Denver Post,* November 24, 2006.

Morgan, Ryan. "Gross Reservoir could double in size under plan." *Boulder Daily Camera,* June 15, 2008.

Neubecker, Kendrick. "The Big Losers: Colorado Rivers." *The Denver Post,* May 22, 2011.

Obmascik, Mark. "Dean of Colorado water lawyers dies." *The Denver Post,* May 3, 1990.

———. "Denver water meters will be mandatory installations planned for 48,000 homes." *The Denver Post,* November 28, 1990.

———. "Last water meter installed in Denver: Program ends years early, under budget." *The Denver Post,* October 30, 1992.

———. "New water chief bridges ideologies: Barry sees self as peacemaker." *The Denver Post,* December 3, 1990.

———. "Water board working at making fewer waves; Two Forks veto brings new 'spirit of cooperation.'" *The Denver Post,* November 8, 1992.

———. "Water folks don't give a dam anymore; Denver board dumps its old logo for a grand, spanking new one." *The Denver Post,* July 25, 1991.

Orwell, George. *1984.* New York: Plume, 1983.

Outdoor Industry Association. "The Outdoor Recreation Economy." http://outdoorindustry.org/images/researchfiles/OIA_Outdoor RecEconomyReport2012.pdf.

Peternell, Drew. "Denver Water sending the wrong message." *The Denver Post,* July 26, 2013.

———. "Thumbs up for Windy Gap water project; thumbs down for Moffat expansion." Guest Commentary. *The Denver Post,* January 10, 2013.

PikaNET: A Citizen Science Monitoring Program For The American Pika. Mountain Studies Institute. www.mountainstudies.org/climate-change-work/2014/11/5/pikanet-a-citizen-science-monitoring-program-for-the-american-pika.

Proctor, Cathy. "Cry heard for storing more water." *Denver Business Journal,* February 23, 2003.

———. "Denver Water Issues RFP for Northern Water Study." *Denver Business Journal,* June 8, 2003.

Russell, Carmen. "Defending a Western River with Art and Collaboration." *National Geographic,* February 12, 2014. http://news.nationalgeographic.com/news/2014/02/140212-fraser-river-colorado-kirk-klancke-water.

Save the Colorado. www.savethecolorado.org.

Snider, Laura. "An 'environmental pool' for South Boulder Creek." *Boulder Daily Camera,* July 10, 2011.

———. "Diverting the Colorado: 2 projects with Boulder County ties to bring more water to Front Range." *Boulder Daily Camera,* July 9, 2011.

———. "Gross Reservoir Expansion Could Restore Stream Flows." *Boulder Daily Camera,* August 9, 2009.

———. "Public Expresses Concern Over Gross Reservoir Expansion." *Boulder Daily Camera,* January 20, 2011.

———. "Public Input Sought on Gross Reservoir Expansion." *Boulder Daily Camera,* November 30, 2009.

———. "Rep. Jared Polis Expresses Concern Over Gross Reservoir Expansion." *Boulder Daily Camera,* February 23, 2010.

Sprout Foundation. www.sprout-foundation.org.

Tapped Out: The Upper Colorado on the Brink. Boulder, CO: The Story Group, 2010. http://thestorygroup.org/uppercolorado.

Tetra Tech, and HabiTech, Inc. *Final Draft Report, 2011 Spawning Bar Core Samples, Grand County, Colorado.* January 31, 2012. http://co.grand.co.us/DocumentCenter/View/3698.

———, HabiTech, Inc., and Walsh Aquatic Consultants, Inc. *Draft Report, Stream Management Plan, Phase 3, Grand County Colorado.* August 2010.

Todd, Andrew S., et al. "Development of New Water Temperature Criteria to Protect Colorado's Fisheries." *American Fisheries Society* 33, no. 9 (September 2008): 428–442.

U.S. Army Corps of Engineers. *Environmental Impact Statement —Moffat Collection System Project.* www.nwo.usace.army.mil/Missions/RegulatoryProgram/Colorado/EISMoffat.aspx.

Vidergar, Cyril. "Footsteps of Ike." *Grand County Living,* Summer 2010.

Willoughby, Scott. "Fraser losing life; Ehlert says diversions damaging food sources on river." *The Denver Post,* March 30, 2011.

Wockner, et al. *Denver Water's Moffat Project FEIS Released—a Lose-Lose Boondoggle.* Press Release. April 22, 2014.

Woodka, Chris. "Low-flow appliance law sought; Denver Water foresees statewide savings of 40,000 acre-feet annually." *The Pueblo Chieftan,* August 24, 2013.

Zaffos, Joshua. "Water projects stuck in regulatory limbo." *BizWest,* September 6, 2013.

Chapter 18: Learning by Doing

Best, Allen. "Alchemy on the Fraser River; Less is More for the Fraser River, Say Parties in Moffat Firming Agreement." *Mountain Town News,* March 26, 2014. http://mountaintownnews.net/2014/03/26/alchemy-on-the-fraser-river.

Blevins, Jason. "The South Platte: A River Untamed." *The Denver Post,* July 10, 2007.

Colorado River Headwaters Chapter, Trout Unlimited. "Learning by Doing." www.coheadwaters.org/ConservationScience/LearningByDoing.aspx.

Colorado Wildlife Federation. "Moffat Project: asks Army Corps in permit to include commitments made by Denver Water." June 10, 2014. http://coloradowildlife.org/our-stand/moffat-project-asks-army-corp-in-permit-to-include-commitments-made-by-denver-water.html.

Denver Water. "Denver Water, Trout Unlimited, Grand County reach agreement on river protections for Moffat Project." March 4, 2014. www.denverwater.org/AboutUs/PressRoom/8C298A85-A0F5-28B5-78FCE2CE3D6041C6.

Grand County Planning Commission. Meeting Minutes, July 11, 2012. http://co.grand.co.us/DocumentCenter/Home/View/1406.

Harden, Mark. "Cover story: Dam deals that wouldn't wash." *Denver Business Journal,* July 4, 2014.

Hickenlooper, John W. Governor of Colorado. Letter to President Barack Obama. June 5, 2012. www.northernwater.org/docs/LatestNews/LettertoPRESIDENTJune52012.pdf.

Land Use History of North America. Biotic Communities of the Colorado Plateau, Riparian Areas. http://cpluhna.nau.edu/Biota/riparian_communities.htm.

Larsen, Leia. "Adapting for river health is key in Moffat Tunnel Collection System agreement." *Sky-Hi News,* May 13, 2014.

Lochhead, Jim, and David Nickum. "Together, we can meet Colorado River challenges." *The Denver Post,* June 1, 2013.

Moffat Collection System Project, Grand County Mitigation and Enhancement Coordination Plan. February 13, 2014. www. coloradotu.org/wp-content/uploads/2014/03/Moffat-MCEP-2-13-14-final.pdf.

Northern Water. *Intergovernmental Agreement for the Learning by Doing Cooperative Effort.* www.northernwater.org/docs/WindyGapFirming/SubdistrictAgreementsWestSlope/WGFP_LBD.pdf.

Osnos, Evan. "Crossing the Water." *The New Yorker,* October 1, 2015.

Trout Unlimited. "Highlights of South Platte Protection Plan." Summer 2001. www.coloradotu.org/wp-content/uploads/2011/05/SPPP-SUMMARY.pdf.

U.S. Congress. Senate. 75th Congress, 1st Session. Document No. 80. *Synopsis of Report on Colorado-Big Thompson Project, Plan of Development and Cost Estimate Prepared by the Bureau of Reclamation, Department of the Interior.* June 15, 1937.

U.S. Department of Agriculture. "The South Platte Protection Plan." www.fs.usda.gov/detail/psicc/landmanagement/planning/?cid=STELPRDB5226762.

Whiting, Mely. "Front Range groups must protect resources." *Grand Junction Sentinel,* Letters, March 7, 2011.

Willoughby, Scott. "Plan for Fraser River is a good one." *The Denver Post,* March 5, 2014.

———. "A victory for conservation on upper Colorado." *The Denver Post,* December 5, 2012.

Chapters 19 and 20

American Rivers. "Why We Remove Dams." www.americanrivers.org/initiatives/dams/why-remove.

Antal Borsa, Adrian, Duncan Carr Agnew, and Daniel R. Cayan. "Ongoing Drought-induced Uplift in the Western United States." *Science* 345, no. 6204 (September 26, 2014): 1587–1590.

Antonacci, Karen. "Chimney Hollow Reservoir gets OK; will add water storage for Longmont." *Longmont Times-Call,* December 19, 2014.

Babbitt, Bruce. *Cities in the Wilderness: A New Vision of Land Use in America.* Island Press, Washington, DC: 2005.

Browne, Malcolm W. "Dams for Water Supply Are Altering Earth's Orbit, Expert Says." *New York Times,* March 3, 1996.

Bushong, Steve. Porzac, Browning and Bushong LLP. Letter to Board of County Commissioners, Grand County. August 23, 2012. http://co.grand.co.us/DocumentCenter/Home/View/1400.

Chouinard, Yvon. "Op-Ed; Tear Down 'Deadbeat' Dams." *New York Times,* May 7, 2014.

Colorado Parks and Wildlife. *Nuisance Wildlife Laws in Colorado.* Updated January 2015. https://cpw.state.co.us/Documents/WildlifeSpecies/LivingWithWildlife/NuisanceWildlife.pdf

The Denver Post. "Big Thompson Flood of 1976." July 31, 2012. http://blogs.denverpost.com/library/2012/07/31/big-thompson-flood-disaster-colorado-1976/2795.

Dischinger, Joseph B. "Local Government Regulation Using 1041 Powers." *The Colorado Lawyer* 34, no. 12 (December 2005).

Beard, Daniel P. *Deadbeat Dams: Why We Should Abolish the U.S. Bureau of Reclamation and Tear Down Glen Canyon Dam.* Boulder, CO: Johnson Books, 2015.

Colorado Department of Local Affairs. 1041 Regulations. www.colorado.gov/pacific/dola/1041-regulations.

Gardner, John. "Breaking news: Northern Water moves forward with Windy Gap Firming Project." *Berthoud Weekly Surveyor,* December 19, 2014.

Grand County. "Windy Gap Firming Project." www.co.grand.co.us/428/Windy-Gap-Firming-Project.

Jaffe, Mark. "Current Affairs on State Water; Utilities Cross the Divide to Start Negotiating Water-moving Plans." *The Denver Post,* October 5, 2008.

Limerick, Patricia Nelson, with Jason L. Hanson. *A Ditch in Time: The City, the West, and Water.* Golden, CO: Fulcrum, 2012.

McComb, David G. *Big Thompson: Profile of a Natural Disaster.* Boulder, CO: Pruett Publishing Company, 1980.

National Oceanic and Atmospheric Administration. "Colorado Remembers Big Thompson Canyon Flash Flood of 1976." www.noaanews.noaa.gov/stories/s688.htm.

Nijhuis, Michelle. "Movement to Take Down Thousands of Dams Goes Mainstream." *National Geographic,* January 29, 2015. http://news.nationalgeographic.com/news/2015/01/150127-white-clay-creek-dam-removal-river-water-environment.

Noreen, Barry. "Another water project is drowned." *High Country News,* December 12, 1994.

Northern Water. "Home and Garden Conservation." www.northernwater.org/WaterConservation/HomeandGardenConser.aspx.

———. *Windy Gap Bypass Funding Agreement*. www.northernwater. org/docs/WindyGapFirming/SubdistrictAgreementsWestSlope/ WGFPBypass_Funding_Agreement.pdf.

———. *Windy Gap Firming Gains Grand County Commission's Approval*. Press release. December 4, 2012. www.northernwater.org/docs/ News_Releases/WindyGapFirmingGains_ReleaseDec42012.pdf.

———. "Windy Gap Firming Project." www.northernwater.org/Water Projects/WGFProjectOverview.aspx.

Reisner, Marc. *Cadillac Desert: The American West and Its Disappearing Water*. New York: Penguin, 1993.

Stengel, Amy. "Water Projects and Colorado's 1041 Regulations." Colorado Riparian Association, September 19, 2009. http:// coloradoriparian.org/water-projects-and-colorados-1041- regulations.

Tetra Tech. "Windy Gap Reservoir Modification Alternative Planning Workshop." Meeting Minutes. Golden, Colorado. September 18, 2013.

———, and HabiTech, Inc. *Final Report: Windy Gap Reservoir Modification Study*. February 2015.

———, HabiTech, Inc., and Walsh Aquatic Consultants, Inc. *Final Draft Report, 2014 Monitoring Report, Grand County Colorado*. February 15, 2015.

Tyson, Neil deGrasse. *Space Chronicles: Facing the Ultimate Frontier*. New York: W. W. Norton, 2012.

U.S. Department of the Interior, Bureau of Reclamation. *Reclamation Signs Record of Decision for Windy Gap Firming Project in North Central Colorado*. Press release. December 19, 2014. www. northernwater.org/docs/News_Releases/Reclamation_%20 release_12-19-14.pdf.

———. *Record of Decision, Windy Gap Firming Project, Final Environmental Impact Statement*. December 19, 2014. www.usbr. gov/gp/ecao/wgfp_feis.

Woods, Mark. *The Holy Cross Wilderness Area and Homestake II: Case History, Conflict Analysis and Intractability*. Conflict Research Consortium, Working Paper 94-64. February 1994. www.colorado. edu/conflict/full_text_search/AllCRCDocs/94-64.htm.

ABOUT THE AUTHOR

STEPHEN GRACE IS THE author of many books, including *Dam Nation: How Water Shaped the West and Will Determine Its Future, Grow: Stories from the Urban Food Movement,* and *The Great Divide.* He served as a consultant for the film *DamNation* and was the scriptwriter and an associate producer for *The Great Divide* film. He lives in the high desert of Colorado.

To follow the progress of the Windy Gap bypass and other efforts to restore the upper Colorado River, please visit the Upper Colorado River Alliance website.

www.UCRA.us